ENGLISH COUNTRY
LIFE & WORK

THE WOODLANDERS.

ENGLISH COUNTRY LIFE AND WORK

RURAL WAYS OF LIFE IN DAYS GONE BY

By

ERNEST C. PULBROOK

(JM)

CLASSIC EDITIONS

PREFACE

To attempt within the compass of a small volume to depict the intricate many-sided life and work of the English Countryside in past and present might well seem a task of almost insuperable difficulty. The subject is so vast and absorbing that a life work would not be too short for it, and enough material could be forthcoming to fill a hundred volumes. The present writer is well aware of the very considerable shortcomings of his effort, but his endeavour has been to present a sketch, as well as he could, of some features of rural life as it has been lived in former generations, and as it exists at present, for those who are interested in Country life in England—a class which includes almost everyone, even under the industrial conditions of town life to which most people are now restricted.

In his earlier work the author dealt with the chief features of the English Countryside. The present is planned as a companion and complement to describe and illustrate the life and work of the human types, of which the landscape forms the scene and setting.

The time did not seem inopportune for a rapid survey of Country Life. After a period of neglect the war has taught the nation the essential pre-eminence of the labours of those who till the soil, and country life is undoubtedly regaining some of the importance and interest it deserves, which it should never have lost. Even though the present period is one of rapid and widespread transition, and the position is again acute in its depression and difficulty, the awakening movement will continue. Much has been written on the different features of the countryside, and on its sport and natural history, but comparatively little attention has been devoted to the countryman and his round of work and leisure. To town dwellers the life of the countryside is almost entirely a sealed book; the neglect of the countryside by politicians is proverbial, and often the countryman is made the butt of flippant pleasantries, or serves as a vehicle for the propagation of some or other set of extreme views.

It has been the writer's desire to record, however imperfectly, something of all the wonderful variety of work on field, farm, hill, common, moor, woodland, and waterside, to glance at country trading and gatherings for sport, commemoration or worship, as well as the features of the past, with particular reference to their influence and survival. In the countryman are rooted the most enduring traditions and best characteristics of the English race, and probably no one realises the incalculable debt we owe to the enterprise and resource of a continuous succession of unknown skilled workers from Neolithic days onwards ; it is they who have made civilisation possible.

Though the author does not desire to lay claim to any striking originality or learning, his work is at least founded on knowledge derived from close acquaintance and familiar intercourse, not untouched with sympathy and feeling. He will be glad if the following pages serve the purpose of a signpost, pointing the way to a region of much attraction, too often neglected. The signpost is merely a directing hand, and cannot mention everything that will be seen on the journey, but leaves the traveller to look round, and observe for himself the variety of features and detail on the way he is traversing. It will also be a source of satisfaction if his work can help to introduce the average Englishman to the everyday life of those who live outside cities, and perhaps awaken a desire for closer acquaintance which may develop into lasting friendship.

The compilation of these chapters has occupied the leisure time of several years, and much assistance and suggestion have been derived from a number of writers, old and new, and from people of many classes and types in the countryside itself. It is not possible adequately to express all indebtedness for this, but particulars will be found, as far as possible, in the Note of Acknowledgment which follows.

September, 1922.

NOTE OF ACKNOWLEDGMENT

*The subject illustrated on the upper part of the page is denoted by " a,"
and the lower subject by " b."*

For permission to reproduce photographs, the author is indebted in many quarters, and in compiling a book of this nature he must express his gratitude to people, often unknown, who have supplied information, obtained during travels about the countryside, as well as to a number of writers, old and new, who have been drawn upon for facts and opinions. Detailed acknowledgment of all these helpers is of course not practicable, but their aid is incorporated somewhat after the analogy of the work of farmer and ploughman in the finished loaf.

Characteristic pictures of English country life and work have not been easy to obtain, and the illustrations reproduced are selected from several hundreds. From Messrs. Frith's excellent series are taken subjects on pages 7, 8b, 10b, 24b, 25b, 58a, 84b, 87a, 91b, 127b, 133a, 134, 156a, 166b, 168b, 177b, 218a. From the extensive and varied collection of Mr. Taunt, of Oxford, come the illustrations on pages 9b, 51a, 66, 133b, 142, 147a, 148b, 149b, 167, 184, 190, 198b, 199. The sons of the late Sir Benjamin J. Stone have kindly permitted the inclusion of the interesting examples on pages 83, 172, 191, 192, 209b, and the Rev. E. A. Godson (Shropshire) has recorded the characters illustrated on pages 3a, 26b, 92, 102b, 155a, 166a, 212b. To the late Mr. W. Pouncy, of Dorchester, author and publisher are indebted for several examples reproduced. The provenance of the other illustrations is as follows :—Mr. Percy Bedford, 36, 52b, 78, 136b, 233a ; Miss M. Billington, 147b; Mr. S. Bridger, Walthamstow, 84a; Mr. Millman Brown, of Shanklin, 8a ; Messrs. Cassell, 87b ; Mr. Cox, of Lincoln, 102a, 121b, 127a ; Mr. Fred. H. Crossley, 1, 17a, 62a, 65, 72a, 88a, 150a ; Mr. Horace Dan, 57a, 97, 98, 122 ; the late Mr. W. Galsworthy Davie, 4a, 18b, 44b, 48a, 51b, 52a, 57b, 58b, 77, 165, 212a, 219a, 219b ; Mr. J. P. Day, Harrow, 168a, 232b ; Mr. W. M. Dodson, 37a, 43a, 150b, 156b ; Mr. S. A. Driver, Ardleigh, 53b, 183a, 210b, 218b ; Mr. Lloyd

vii

Elias, of Llandudno, 101a, 104b, 225 ; Mr. Forder, of High Wycombe, 226a ; Mr. W. R. Gay, of South Brent, 10a, 101b ; Mr. Goulding, of Beverley, 13, 14 ; Mr. Benjamin Hanley, of Selby, 17b, 62b, 178b ; Messrs. Hayman and Son, 23 ; Mr. George Hepworth, Brighouse, 37b, 38b, 48b, 62b, 104a, 189a, 210a ; the Rev. Hugo Heynes, 135, 189b ; Messrs. F. R. Hinkins and Sons, 120, 121a, 198a ; Mr. Alex. Old, of Padstow, 197 ; " The Oxford Journal," 200, 209a, 211 ; Mr. P. Perrin, of Conway, 44a ; the Photochrom Co., 71b, 103a, 113, 141, 155b ; Mr. C. Reid, of Wishaw, 182a ; Mr. J. H. Spittles, Gt. Missenden, 119b ; Mr. Stoate, 47a ; Messrs. J. Valentine, 178a ; Mr. Vasey, of Abingdon, 9a, 54 ; Mr. Henry Walker, Louth, 18a, 88b, 181b ; Mr. J. C. Warburg, 171a ; Mr. W. F. Wastell, frontispiece, 3b, 4b, 61, 71a, 103b, 177a ; Messrs. Wilkinson, of Trowbridge, 217a. The subjects illustrated on pages 35b, 53a, 128a, 233b, are reproduced from the author's own photographs.

In compiling this note every possible care has been taken, but with the large amount of material handled, extending over a number of years, it is always possible that the acknowledgment of some contribution may have been omitted, or wrongly attributed, and if there is any mistake of this character it is hoped that it may be excused.

CONTENTS

CHAP.		PAGE
I.	THE COUNTRYMAN AND THE NATION	I
II.	THE SQUIRE AND HIS FRIENDS	22
III.	FARMERS, OLD AND NEW.	39
IV.	FIELD WORK	55
V.	COTTAGE FOLK	70
VI.	THE VILLAGE CRAFTSMAN	85
VII.	AMONG THE HILLS AND ON THE COMMONS	100
VIII.	IN THE WOODS	112
IX.	CALLINGS OF THE COAST AND WATERSIDE	126
X.	ON THE ROAD AND IN THE LANE	143
XI.	TRADING IN THE COUNTRY	158
XII.	THE COUNTRYMAN AND HIS LEISURE	173
XIII.	OLD-TIME CUSTOMS AND FOLK LORE	188
XIV.	RELIGIOUS LIFE—PAST AND PRESENT	205
XV.	WAR TIME AND ITS AFTERMATH	222
	INDEX	237

THE VILLAGE BLACKSMITH.

I.

THE COUNTRYMAN AND THE NATION.

" I think the extraordinary blessings and privileges of English rural life have never been sufficiently considered. It is only when we begin to count them up that we become aware of their amount, and surpassing character."—William Howitt.

IF you would understand the soul of England you must wander far in the highways and by-ways of the country-side. Her energy, her enterprise, her industry will be found in the cities and the towns, her painstaking labour, her plodding patience, her latent strength in the country. " The stranger who would form a correct opinion of the English character must not confine his observations to the Metropolis. He must go forth into the country ; he must sojourn in villages and hamlets ; he must visit castles, villas, farm houses, cottages ; he must wander through parks and gardens ; along hedges and green lanes ; he must loiter about country churches ; attend wakes and fairs, and other rural festivals ; and cope with the people in all their conditions, and all their habits and humours." So wrote Washington Irving a century ago, and allowing for change of habit and custom and circumstance it remains true to-day. So long as this continues, and only so long, we may look towards the future of England without misgiving. The industries of the city bring us wealth, the work of the country strength and health and self-reliance. If rural England be prosperous and contented the nation will flourish and be free.

The country and its daily tasks have ever inspired the poet and provided texts for the orator who would uplift the people ; the quiet of the fields and woods soothes those vexed by the worries of the city ; and the labour of the unknown and humble through countless generations shows what ceaseless effort and self-effacing co-operation can accomplish. The town supplies the luxuries of life, the country its necessities. Spiritually and materially, a large and flourishing rural population is necessary to the country that would be great. Yet this has been forgotten, and even after a devastating war many still regard agricultural England as of small account, as if all the wisdom and virtue and enterprise belonged to the industrial centres. The country is not merely the background of the city but its complement ; one cannot exist without the other.

In these days, it is not easy to recognise the great part played by the country-side in developing England. We owe the foundation of our prosperity to our sheep, for wool built up first our commerce and then our industries. Clothiers and ship-owners, whose counting houses over-looked the green fields, joined with the wealthy merchants of London in founding those companies whose explorers discovered new lands across the oceans and extended our trade abroad. From mansion and smaller manor house, from parsonage and farm, came statesmen and law givers, commanders on land and sea ; from cottage and hamlet, preacher and writer and reformer who helped to shape our character. Our fleets were manned by the coast villagers, the bulk of our army was recruited from the sturdy country lads and officered by the county families. Thanks to the fact that we were almost entirely self-supporting we held out during the long wars with France, and during the fateful years from 1914–1918 we looked to farmer and labourer to second our sailors in defeating the submarine war.

In the past the country was England. Population was more evenly distributed ; there was no great difference between the agricultural and industrial districts as there is to-day. A century ago those living under rural conditions outnumbered the dwellers in the towns ; in 1851 they were equally divided, but in 1911 the country accounted for only some 22 per cent. True, the same census showed a greater rate of increase for the country than the city, but it represented a faster growth of suburbs and a greater development of mining and industrial areas, thus really an extension of the town at the expense of the country. According to the first returns of the new census the ratio of urban popula-tion still increases. Agriculture still employs more workers than any other calling, but the percentage is much smaller than in 1801, when those dependent on farming and its auxiliary trades amounted to 37 per cent. of the population as against about 10 to-day. In 1906, a Government return showed that between 1881 and 1901 farm labourers had decreased in every county except Anglesea and Flint, although since then the decline has been arrested somewhat.

The chief hope for the future lies in the fact that love of the country is ingrained in the English race, that a country life is the best the world can offer, that eventual retirement to a cottage is the flame that keeps ambition alive in the hearts of thousands of tired workers. Perhaps it is only in the country and sleepy towns that the real England is to be found, for together they form the nursery of the nation. As the child shows inherited characteristics more clearly than the man, so the

THE GARDENER.

AN AGED BUT ACTIVE WORKER.

A WEIGHTY QUESTION—"WILL 'E GI'E PLENTY O' ZIDER, THEN?"

GOOD HEALTH, MATE!

lost in a crowd he lives largely in the public eye, and therefore in an outspoken community must conform to the general idea of what is seemly. He is as unaffectedly interested in his neighbour as he expects his neighbour to be in him, and is a firm believer in the adage that the proper study of mankind is man. The labourer may be uncultivated and perhaps coarse in speech, but as often as not, he posessess natural politeness and kindliness of heart. This is more marked, perhaps, in the hills and among the fens and marshes, where folk are only now being influenced by improved communication with the outer world and retain characteristics less marked elsewhere, being conservative in habits, independent in character, and usually speaking a dialect which has changed but little for generations.

However, the countryman is so elusive a being, so full of contradictions and the unexpected, that he is not easy to describe. Local variations of feature and complexion and build are just as fascinating to study as character. These diversities of type not only furnish examples of the various races from which the modern Englishman is sprung but provide opportunities of noticing modifications caused by environment and occupation, and also, it must be remarked, for drawing false conclusions. In some districts there are individuals—maybe only one or two, maybe a pocket of the original inhabitants left isolated by a wave of immigration—who have little in common with their fellows, sometimes, it would seem being typical representatives of that long forgotten people who raised some of our oldest monuments. It is not always wise to believe the local story that the village or hamlet is occupied by descendants of survivors washed ashore from a ship of the Armada, though very occasionally it may contain a germ of truth, but in places are unmistakable traces of Flemish immigration, more noticeable in name and local words than feature.

One comes across small colonies of folk who have migrated from another district and retaining their characteristics are still regarded as " foreigners." The countryman considers his own people the salt of the earth and his native place the hub of the universe. The more isolated a community, the more clannish and locally patriotic it remains. The Yorkshireman, especially he of the hills and dales, the Cornishman, and the Devonian are examples, while did not the dwellers in Romney Marsh speak about " Romney Marsh, England, and the rest of the world " ?   An old Cornish woman returning home after passing three or four months in the Isle of Wight remarked, " Dear life ! I was real glad to get back, as bein' there was like livin' in a desert."

lost in a crowd he lives largely in the public eye, and therefore in an outspoken community must conform to the general idea of what is seemly. He is as unaffectedly interested in his neighbour as he expects his neighbour to be in him, and is a firm believer in the adage that the proper study of mankind is man. The labourer may be uncultivated and perhaps coarse in speech, but as often as not, he posessess natural politeness and kindliness of heart. This is more marked, perhaps, in the hills and among the fens and marshes, where folk are only now being influenced by improved communication with the outer world and retain characteristics less marked elsewhere, being conservative in habits, independent in character, and usually speaking a dialect which has changed but little for generations.

However, the countryman is so elusive a being, so full of contradictions and the unexpected, that he is not easy to describe. Local variations of feature and complexion and build are just as fascinating to study as character. These diversities of type not only furnish examples of the various races from which the modern Englishman is sprung but provide opportunities of noticing modifications caused by environment and occupation, and also, it must be remarked, for drawing false conclusions. In some districts there are individuals—maybe only one or two, maybe a pocket of the original inhabitants left isolated by a wave of immigration—who have little in common with their fellows, sometimes, it would seem being typical representatives of that long forgotten people who raised some of our oldest monuments. It is not always wise to believe the local story that the village or hamlet is occupied by descendants of survivors washed ashore from a ship of the Armada, though very occasionally it may contain a germ of truth, but in places are unmistakable traces of Flemish immigration, more noticeable in name and local words than feature.

One comes across small colonies of folk who have migrated from another district and retaining their characteristics are still regarded as " foreigners." The countryman considers his own people the salt of the earth and his native place the hub of the universe. The more isolated a community, the more clannish and locally patriotic it remains. The Yorkshireman, especially he of the hills and dales, the Cornishman, and the Devonian are examples, while did not the dwellers in Romney Marsh speak about " Romney Marsh, England, and the rest of the world " ?   An old Cornish woman returning home after passing three or four months in the Isle of Wight remarked, " Dear life ! I was real glad to get back, as bein' there was like livin' in a desert."

THE QUIET OLD-WORLD TOWN—THAXTED, ESSEX

INDUSTRY IN ARCADY—COALBROOKDALE.

HAYSEL—OLD STYLE.

A BUCKINGHAMSHIRE FARM YARD.

THE RUINED LOCK—ONE OF THE EVIDENCES OF RURAL DECAY.

THE RICK FIRE—A COUNTRY EXCITEMENT.

CARTING BRACKEN. THE MOORMAN'S AUTUMN HARVEST.

THE GIPSIES' HOME.

To understand the countryman and his life and to sympathise with his difficulties it is above all necessary to remember that the most marked characteristic of England is her infinite variety. Not only the people themselves but the conditions of life, methods of work, industries, customs, and climate differ from county to county, almost from parish to parish. No one stereotyped plan of rural reconstruction would therefore be suitable ; what would restore prosperity to one place would bring certain ruin to another Consider the diversity of agriculture alone. Here climate and soil point to the growing of wheat, there stock raising and dairy farming are the proper branches to follow. The hop gardens of Kent, the fruit culture of the vale of Evesham and Cambridge, the apple orchards of Devon, are all characteristic of their districts, and the tally might be increased with ease.

Methods and implements are equally diverse, occasioned by soil and contour of the land, so that in one place an unusual method or an archaic system or ancient tool long supplanted elsewhere survives all innovations. On the sheep runs of the north-western fells the flocks belong to the landlord not the tenant ; in Dorset, the dairyman often hires his cows from the farmer who feeds and houses them. Not only do the breeds of domestic animals still largely represent their districts but their harness and trappings are not the same ; even the build of the carts and waggons, the very colours in which they are painted, follow local custom. In fact the diversities of rural England are endless.

Think also of the callings which minister to her wants on the hills and in the vales, along the coast and by the river, in the woods and on the plains, providing the necessities of the town as well as country. Agriculture alone gives employment to thousands who do not till the soil—the smith, the wheelwright, the thatcher, the hedge carpenter, and a score of others. Each occupation acts and re-acts on another, however unconnected they may appear. The depression of one throws the business of two or three others out of gear ; the unexpected death of a wheelwright may delay the harvest of a farmer waiting for waggons to be repaired or wheels bound, the closing of a forge adds to the expenses of the carrier or robs the cottager of the profit on a parcel of perishable goods because horses must be sent away to be shod.

Another factor which has so large an influence on country life is the thread of continuity which runs through it, joining the remote past with to-day. The lover of the country-side recognises it in the features of some well-known landscape—the camp upon the down, the ruined keep that once guarded a pass between the hills, the ancient church, the Tudor manor house, the old cottage—but the landmarks of rural life, the survivals which determine habits and methods of work, are revealed only to the few, to those with some knowledge of the practice of to-day and more than a passing acquaintance with the conditions of other times. The changes have been great and far-reaching, but upheavals have been few, and there has been no sudden break with the past ; whether these changes have been comparatively rapid or very slow the new has always been grafted on the old. How could it well be otherwise ? Few of the arts and crafts of the country-side can be ordered by fixed rules or scientific formulas ; varying seasons and differing local conditions are responsible for almost countless exceptions.

Modern agricultural practice has been slowly evolved through the ages by the skilled observation and every day experience of thousands labouring under dissimilar conditions. Working as the cultivator may all his life in one place, as did perhaps his father and grandfather before him, the unconscious intuition which comes of long practice often makes him the equal of the most scientifically trained worker who has all his local knowledge to gain. When rural education blends scientific knowledge with inherited instinct it will accomplish much, but few can maintain that its achievements have been remarkable so far. Compulsory schooling has not proved better at forming character or imparting a love of craftsmanship than practical training at an early age or the simple teaching of the old-fashioned pedagogue. It tends to deaden observation and weaken the inherited folk-lore that together built up what might be termed the olden peasant culture, which often makes a chance conversation with the passing generation such a delight.

So, too, we find greater contentment and a keener interest in their daily tasks among the old. The veteran hedger deplores the fact that few young men have the patience to learn to lay a hedge, the grey-haired ringer mourns because the present generation prefers to read the evening paper to the joys of mastering the mysteries of Bobs and Singles and Grandsire Triples. For love of good craftsmanship, whether it be displayed in ploughing a field or building a waggon, we must usually go to the older men ; the younger seem to regard their work as an irksome task to be finished and forgotten as quickly as possible. Not that this can be laid entirely upon the shoulders of rural education as it is the tendency of the age, but the elementary school system cramps the good teacher and moulds the majority of the pupils to an indifferent pattern, and, in conjunction with ease of intercourse, is slowly, but surely, stereotyping character and destroying individuality, so strong a feature of the old countryman.

SHEEP-WASHING IN YORKSHIRE.

TEAMS RETURNING HOMEWARD—YORKSHIRE.

The failure to arouse in the young an interest in rural things is one of the causes of the drift to the towns. A desire for higher wages and a craving for amusement have had their influence, but the long stagnation of the country has been more potent. Now the leaven of unrest and discontent is working in the hamlet as in the city, and there is that vague longing for a change so typical of the times. Perhaps the increase of wages will keep more on the land, but it is too early to judge the result or estimate exactly the improvement it will bring. When customs and conditions and weather are allowed for, it may be found that the increase means very little.

As for amusement the country is by no means the dull place it is supposed to be. There is no clear cut division between work and play ; one melts imperceptibly into the other, although here again increased wages and shorter hours are already tending to put an end to the old easy going ways. Wages and methods adjusted to the niceties of scientific economics do not help to sweeten labour or oil the intercourse of master and man. The countryman is an adept at amusing himself ; the village fête, the local race meeting or agricultural show, the ploughing competition, the whist drive, the dance, and the rest, are more to him than the theatre or exhibition to the suburb. A rick fire is an excitement for all, old and young, and there is always the inn, the square, or the cross-roads where the weighty affairs of state, the character of passing farmer, or appearance of casual stranger can be discussed.

Some are led away by the attractions of the city but nothing will induce the true countryman to take up his abode there, however, alluring the prospect, and many compelled by circumstances go with infinite regret. " I should choke in the town," said a prosperous farmer who thoroughly enjoyed his visits to his son in London but clung to the holding on which he had worked all his life. A cottage woman dwelling solitary in an isolated wooded cleft, after stating her nearest neighbour lived nearly a mile away, said merrily, " But there, I don't like London ; why, you always have to have the door to ! " as if living with an open door was the acme of happiness in life. " The prettiest spot in the world," mused a labourer, as, returning home across the down, he stopped to drink in the view, a jutting chalk headland with a fishing village nestling cosily within its flank. Such appreciation may not be so common as it was, but it often comes with regretful longing to the migrant who would return to his native place but seeks an opening in vain. Lack of housing accommodation drives some to the towns, as in the times of depression hundreds of stout old cottages were allowed to decay.

Building bye-laws unnecessarily stringent for country use, the renting of cottages as holiday homes, and, near the towns, the artisan with a cycle, have all had some influence on the shortage of rural houses, but the question is a large one that cannot be discussed apart from economics and politics.

It is often urged that country life is snobbish, that cliques abound, that there is no mingling of the classes, that the poor are neglected and their lives sordid. He who puts this forward as a typical representation of country life must be very unhappy in his experiences and restricted in his acquaintanceships. It may be true of those sordid industrial settlements, neither town nor country, slums disfiguring fair landscapes, which possess the ugly characteristics of both with the advantage of neither. Rural England is certainly not a sunny corner in Arcady where it is ever afternoon and the shadows of life unknown. Country life, like everything else, is a blend of the good and bad, but the good usually tips the scale.

Cliques there may be, but those of like tastes and temperament are few and naturally cling together. Snobs there may be, also their opposite, small shop-keepers or cottagers who mistake rudeness for independence. Class does not melt imperceptibly into class as in the towns because each grade is seldom represented, and the villager hates anything which to him savours of interference or patronage. Unwritten laws determine the ranks of rural society, laws which to casual visitor or new immigrant may seem foolish and snobbish, but residence usually proves them, narrow as they may be in some ways, to have been evolved to suit their own purpose and to smooth everyday intercourse. Yet, as a general rule, real democracy exists in the country ; every one has his recognised place and his definite work. The child of master and man attend the same school ; squire and labourer sit together on the parish council ; the lady from the great house and the shop-keeper serve on the same committee.

To generalise is foolish, but broadly speaking there is less grinding poverty in the country, the real country, than the town. It is an unhappy village which possesses no one to help the unfortunate, and numerous are village charities. It is a tragedy that many old people of independent spirit, dwelling solitary and apart, end their lives in the infirmary, for it is usually the independent workers, the old man who earned his living in the curious ways open to the country and the woman who would be tied to no one task or employer, who come to the workhouse because they have outlived strength and savings. Unless death or misfortune

THE PLOUGHMAN BEGINS HIS TOILSOME DAY.

THE GAMEKEEPER'S GIBBET.

18

INDISPENSABLE IN THE FENS—THE DITCH-CLEANER

SMALL PEDDLING BARGAINS.

overtake a former employer or his family move away, a faithful labourer is not often deserted in his old age. A cottage ill-kept and cluttered with apparent rubbish does not necessarily reflect the poverty-stricken state of its unkempt owner, but may disguise comparative affluence.

Drunkenness and crime bring some to want, but the really vicious move on or go to the town where their record is unknown. The un-written laws of the country make it uncomfortable for the habitual wrong-doer unless he belong to one of those colonies of bad characters found in some villages and small towns. It is a libel to charge the country with wholesale immorality, for though morals may be primitive in some ways there is generally a pretty strict code of virtue which has scant sympathy with transgressors. The ducking stool and the bridle have passed with a more barbarous age, but " rough music " often expresses public opinion, and disapprobation can be shown in other ways.

The wealthy landowner—by no means synonymous words—may occasionally forget his responsibilities and duties ; and the farmer and shop-keeper, especially those who by industry and self-denial have attained some meed of prosperity, may be hard and narrow in their outlook, but against this must be put the cottager who shows a disregard for the convenience of his employer and presumes on the forbearance of those above him. Superstition has by no means died out, though only those intimate with country life can discover the exact facts, and the chemist is usually consulted more than the doctor who is called in only when absolutely necessary. The district nurse, still a new-comer in many parts, is slowly but surely winning her way against opposition and becoming the friend of all, but prejudice and ignorance die hard and modern treatment is still regarded as cruel indifference.

Just now the country is in a state of flux, and those who know it best are least inclined to predict the future. Few of the present problems are really new. The sale of large estates, the decay of villages, the demand for higher wages, easier conditions and more houses, have been witnessed before. Acts of Parliament have been passed to prevent farm being added to farm or ploughland turned to pasture, and to authorize the building of cottages with a decent plot of land attached. The causes and results of these may be studied with advantage if we are to turn the present awakening to account. Let us remember that life in the country has always been largely self-contained, and though it is still often said contemptuously that people exist by taking in each other's washing such mutual support leads to independence and usually a modest competence and so contentment, unless conditions being too hard life becomes a never ending struggle for mere subsistence.

Rural England has long been under a cloud, but every cloud passes in time, and it looks as if the recent explosion may have dispersed that which for so long has darkened country life. However, a sudden burst of sunshine has its disadvantages ; it dazzles vision and often causes the inessential to stand forth with undue prominence. Only by shading the eyes from the glare and studying the prospect with discerning gaze can a true impression be obtained. To recognise the importance of a prosperous and contented country-side will be vain if the fundamental facts that govern its welfare are not understood. Good intentions unaccompanied by knowledge and experience may work more harm than studied neglect. Rural England, thanks to its tenacity and its patience and its industry, worked out its salvation in the face of neglect but it would be powerless to restore what might be destroyed. Let us not, therefore, be over hasty in reforming zeal or endeavour to rebuild before learning the reasons which caused the conditions now condemned.

Flourishing agriculture would indeed transfigure the face of England, for when it thrives a host of other trades and industries flourish in its train, and a reserve of sturdy manhood is raised for town and Dominion. New careers will be opened to those who long to shake the dust of the city from their feet and live more in the open air. New industries will arise to supply local needs, more money will be kept at home, and cultivators certain of their profits will be able to experiment for themselves with less haunting fear of finding only ruin. The country is the home of the craftsman, the man who works leisurely and lovingly, thinking as much of the work as the payment it brings. Even the farm labourer is a skilled workman whose value is measured by his experience ; he is not a human machine who accomplishes one process over and over again until the brain grows dizzy from repetition and the mind is hypnotised by discontent. True, he does the same work from year to year, but conditions always vary so that his knowledge and his skill are brought into use. Above all, he not only sows but reaps ; he sees the beginning and the end of his task and he knows if it has been well accomplished.

Tilling the soil is one of the oldest of crafts and remains one of the most satisfying. It strengthens the muscles ; it soothes the mind ; it means the pitting of man's intelligence against the vagaries of Nature, and when difficulties have been overcome and victory achieved it is a bloodless victory that brings gain to all mankind. Agriculture is more than an industry which can be improved by the application of science and the use of machinery ; it is a nursery for the human race, an occupation that will provide a healthy stock to keep the nation from decay, a buttress against unrest, and the path by which simple contentment may be reached. Lastly, in times of peril, it provides with the Navy the twin sheet anchor which alone will prevent the country from drifting on the rocks. Well will it be for the nation if it remembers the words of Cicero :—" Of all things from which any gain is obtained, there is nothing better than agriculture, nothing more productive, nothing sweeter, nothing more worthy of a man, or of one who is free."

II

THE SQUIRE AND HIS FRIENDS.

ENGLAND owes much to her landed gentry, the nobles whose parks add so much distinction to her landscapes, and the lesser squires whose houses may be found in almost every village. Their work for the country is inscribed on countless pages of her history, the benefit they have conferred on their own neighbourhoods is revealed by a study of the local records or perhaps by a chance conversation with an aged farmer who tells how he was helped over bad times. Yet the wicked and corrupt have given their order a reputation it does not deserve, for the deeds of the notorious are handed down in ballad and story while the work that makes for the good of the community is forgotten by the succeeding generation.

The tradition that the great landowners have held their manors for centuries is mainly a tradition alone. From the beginning our country estates have constantly changed hands. Consider the revolts and civil wars, great and small, when the lands of the losing side were confiscated ; think of the many battles in times when serious wounds nearly always meant death ; remember the epidemics which carried off peer and peasant impartially when both were equally ignorant of hygiene. Where are all those great and wealthy families which flourished in Elizabeth's day ? They were prolific, their manors many, yet how few exist to day, at least in anything like their former glory. One generation sometimes seems to have exhausted itself by its fruitfulness, for the next had few or no children and the heir being killed in the wars the estates passed with the daughters.

Sometimes local fame had disastrous results, the Champernownes being an example if the story be true. In Tudor days one of them took pride in his band of musicians with which he made a brave show at the court of Henry VIII, but finding the visit very expensive vowed never to repeat it. Later Queen Bess invited him to town with his musicians but remembering his former experience he declined on the plea of lack of means. " I'll impoverish him still more," declared the enraged lady and found reason to compel him to part with more of his manors. An idle tale maybe, but it helps to explain the decline of once

A COUNTRY HOUSE PAGEANT: AN ELIZABETHAN MASQUE AT SYDENHAM HOUSE, CORYTON, DEVON.

BOYTON MANOR, WILTSHIRE.

BOURNE HOUSE, BRIDGE, KENT.

HEVENINGHAM HALL, SUFFOLK.

MADELEY—OLD COURT HOUSE.

AN ELIZABETHAN ROOM, BADMINTON.

THE HOUNDS MEET BEFORE THE HOUSE.

well-known families. In a building age some squires aspired to possess a home worthy of the times and after designing a mansion beyond their means were obliged to sell to retrieve their fortunes.

The Civil War was responsible for the ruin of families who stood for the King ; at the Restoration they recovered their land but little else, and struggling under a load of debt sold their estates to regain some of the money expended in the Royal cause. In the eighteenth century extravagance caused the downfall of many landowners ; they kept hounds and fighting cocks, raced and wagered, were patrons of the rough sports of the day, and maintained a house in town. The hard times which followed the Napoleonic wars, and more recently the decay of agriculture helped to thin the ranks of less wealthy squires, especially as the more shrewd among them found better openings in business than the land.

Lovers of a settled order are dismayed by the numbers of estates changing hands, and sentimentalists deplore the new blood brought in to the circle of landowners. Neither complaint is new ; both figure over and over again in the writings of old chroniclers, in the entries of diarists, and in the pages of the romances of the past two centuries. It has been said that between the reign of Charles II and Queen Anne no fewer than 585 private Acts of Parliament for the sale of lands were passed, and in 1695 Evelyn entered " Never so many private Bills passed for the sale of estates, shewing the wonderful prodigality and decay of families." Then there is the oft repeated regret that the aristocracy of the land is being replaced by wealthy kings of industry. But England did not sneer at trade until taught to do so by imported ideas. When the royal and warrior aristocracy came largely to an end it was supplanted by one of trade. What were the nobles and knights of Elizabeth's day but merchant adventurers who traded and fought as occasion offered ? Elizabeth herself was the descendant of a mercer's apprentice from Sall, in Norfolk, who became successful and was made Lord Mayor of London.

Country and town were so closely connected that fortunes founded on agriculture were increased by judicious speculation in trading ventures. Most of the greatest and oldest families had junior members engaged in agriculture or industry, and the smaller landed gentry were nearly always interested in commerce. Not at all wealthy as we understand the word to-day, they were largely farmers by profession and apprenticed their sons to trade. Their children frequently made quite humble marriages, and their descendants may be found among the shop-keepers of country towns or in callings even less aristocratic, though name must not always be relied on, for at one time it was quite common for servants to take the names of their masters. Many were wool and cloth merchants, and those in the maritime counties turned their attention to shipping.

Patriotic to the core, they were always ready to place their persons and their fortunes at the disposal of sovereign or country, and from their ranks were recruited the commanders of expeditions which opened new trade routes, explored unknown seas and countries, or " singed the beard " of the King of Spain. In the records of every shire and town you will find their names though most have been forgotten. They were not always rewarded for their services ; kings have ever possessed little gratitude, and the nation too frequently forgets those who have come forward in hours of peril. George Raleigh, half-brother to the famous Sir Walter, was landowner, farmer, miller, and merchant ; when he provided a ship for the fleet that fought the Armada, the city fathers of Exeter were so slow in making payment that Sir Francis Walsingham admonished them for their " slackness." Even the word is familiar.

Although so many of the old squires' families have passed away their monuments remain in the church, perhaps their charities still enrich the village, and scores, if not hundreds, of their manor houses attract attention—maybe a large farmhouse partly in ruins, maybe a couple of labourers' cottages formed in the wing that alone represents the original building. Some, like Madeley Court in Shropshire, have fallen to ruin, but neglect and decay cannot hide their distinction, and within recent years many have been restored to something like their old condition and have become once more the homes of well-to-do families. However, large numbers have never fallen from their original position; well kept and surrounded by their neatly clipped hedges, terraced gardens, and small parks, they remain one of the most typical features of the country-side.

Sometimes manor house and church stand close together with the village under their shadow, a reminder of the old country life that is rapidly passing away, whilst others occupied as farms are often remote from the highway at the end of a track that can be called a road only by courtesy, although some are approached along avenues of aged trees with gaps here and there marking the ravages of winter storms. Sometimes they withdraw modestly from public gaze behind thick hedge or high wall, showing only a gable or a barn that was once a chapel, or they may face the road for all to see. The projecting porch bears the escutcheon or initials of the original owner, perhaps with the date of erection or a motto, and the massive door is in keeping with the mullioned windows. They may incorporate portions of a still older building, perhaps once a fortified house dating from the days when private wars and feuds could not be disregarded, or are surrounded by a moat, and a few, such as Stokesay Castle, are landmarks in the development of the English house. In others the old hall yet remains open to the roof, while great numbers show panelled rooms, finely carved chimney pieces, moulded ceilings, and maybe a frieze of coats-of-arms depicting the alliances of the family.

These old manor houses are part of the texture of the life story of England. In that upper chamber a famous man first saw the light ; within the walls of that secluded dwelling was held a Council that had far reaching effects for good or ill ; with another is connected nursery rhyme or story repeated by a mother to her child. Here are secret hiding places or priest holes ; there ancient customs are observed ; elsewhere grim relics are cherished, such as the skull in a Somerset farm-house supposed to bring disaster to all who attempt its burial. As for ghosts their name is legion—hunting squires who return at midnight to the scene of their former exploits, spectral coaches that roll silently up the drive before the death of an inmate, apparitions that give warning of peril, to which might be added omens attached to certain families.

A chance conversation often reveals how these legends and stories persist. " To the giant's house " was a labourer's reply to a question concerning a footpath in a quiet corner of Sussex. This is Brede Place, that pretty manor house set amidst trees, so called to this day because the story goes that one of the earlier owners was an ogre who ate children. In a nook of the Brendon Hills in Somerset is the estate of Combe Sydenham which rejoices in both ghost and legend. Down the combe at midnight comes riding the ghost of Sir George Sydenham, a notable Cavalier, and in the house is the cannon ball which Drake fired to warn Elizabeth Sydenham, about to marry an importunate suitor, that he was very much alive. Much the same tale is told of Saltash and his first wife, but Combe Sydenham can produce the identical cannon ball which according to local gossip cannot be taken away. Not so long ago it was removed, and behold ! shortly afterwards it was back in its old place.

These smaller mansions are more pleasing to the eye and more thoroughly native than the huge porticced piles built in later days, their design largely influenced by foreign fashions and their decorations often carried out by artists and craftsmen from other lands. Usually, typical of their age, the larger buildings are certainly characteristic of rural

England, but they do not seem to form part of the landscape as do the smaller houses and cottages.   The parks which surround them and the scenes in which they are set belong to England alone, but a distant peep is nearly always preferable to closer inspection, when they become fine examples of the architect's art and witnesses of their owner's opulence, and are buildings rather than homes as understood by the average Englishman.   Some are distinctly ugly, and it has been said that one landowner returning to gaze on the Georgian mansion erected during his absence in the place of his three centuries old home declared " Well, they have succeeded in building me the ugliest house in England." In this respect, at least, the modern generation shows better taste and greater wisdom by taking purely English types as models and building country houses in our native style.   Modern Tudor may be somewhat staring until its newness is toned down by weather, but it does not offend the eye as do some of the castellated structures of the pretentious.

Sometimes beside the manor house is an ancient barn of huge dimensions or a large pigeon cote, a reminder of the days when winter feed being difficult to obtain most of the animals were killed and salted down, pigeons providing almost the only fresh meat.   Of those days when the great families had many estates and travelled from one to another it has been said that the nobles ate their way round the country. Pretty well all old family and manorial records contain orders to stewards or bailiffs to prepare for the coming of the Master, to send a buck from the park, and so on.   When the modern squire makes pilgrimage from one country house to another it is for the shooting or the hunting, and what the estate cannot provide is forwarded from town, but the local shop-keepers are also busy fulfilling the list of orders that may be expected on such occasions.

Times may have changed but the life of the country squire, great and small, is the finest the world has to offer.   Nor is it cause for wonder that the pleasures too frequently hide all the hard work that occupies the hours between.   These great houses, and many of the smaller ones, possess all that makes life worth living ; fine libraries and noble picture galleries ;   records and collections for the scholar and student ; extensive and well laid out gardens, and range of greenhouses where exotics bloom ; secluded nooks and quiet reaches of stream or lake where the nature lover can pass the time undisturbed ;   and sports and pastimes to suit the young and old, energetic and lethargic.   The hounds meet periodically upon the drive before the entrance ;   the well stocked coverts, noted for their difficult shots, attract the gunner in the proper season ;   trout streams offer their quieter pleasures ; but to the regret of many, golf seems to be ousting country house cricket. A corner of the park is used for polo or is the scene of the local hunt's gymkhana, the lawn behind the house forms an ideal setting for a pageant to swell the funds of a county charity, and periodically the gardens are thrown open for political fêtes.

The births and marriages of long settled families are seldom mere private matters but concern everyone. The tenantry take part in the festivities and there are great doings in the village, as at Wentworth Woodhouse, a few years ago, when the heir to the Fitzwilliams was christened and an ox was roasted whole, not that that was sufficient for the 7,000 guests invited. Few celebrations are on this huge scale, but there cannot be many villages where such events are not observed by the lord of the manor and tenants in some shape or form, if only by the merry peal on the bells and the beflagged street. The tenants' present to the heir on his coming of age, or daughter on her marriage, is proudly displayed in one of the local shops. When the old squire dies the village laments, blinds are drawn, people talk in subdued tones, the bell tolls, and a knell is rung.

Regret at his passing is seldom confined to his estate and village, for he will almost surely have been a member of many public bodies. As a rule the less his name appears in the newspapers the greater his service to his own district ; probably he prefers to follow the old tradition and be the father of his people so far as modern conditions allow. Country folk may be proud of the man of affairs whose name is a household word throughout the land, but they have an affectionate regard for the landowner who dwells among them and concerns himself with their immediate interests. He is chairman of half a dozen local bodies, and the usually despised Parish Council may have at its head the local squire, be he a great landed proprietor or merely the possessor of the manor house and a few acres. Feudal times have passed away but the landowner still has many dependents though these grow fewer every year ; he does not forget the poor and the local hospital receives gifts of money and in kind at stated seasons ; he may seldom follow the old custom of founding almshouses but that is because he dispenses his charity in other ways.

He interests himself in agriculture and, mayhap, breeds pedigree stock or runs a model farm where experiments in cultivation are carried on, the best known example of the latter being that at Rothamsted, founded by Sir John Bennet Lawes. He takes the lead in local enterprise and sets the pace in times of emergency, and his influence on the neighbourhood is incalculable. Where he keeps his farms and houses in good repair, seeing the land is well and cleanly cultivated, the smaller landowners follow his example and enterprise is not wanting ; where he is an invalid unable to take an active share in the work of life, an absentee little interested in country pursuits, or the possessor of an overburdened estate, stagnation and apathy hang over the land. Our great landlords are blamed for not running their estates on business lines, but if they leave the management in the hands of a hard headed agent it is usually declared they fail to realise their responsibilities. Where many small landowners are found, cohesion and local solidarity are usually missing unless among them is a man of outstanding character.

Modern squires are frequently successful business men who seldom spend sufficient time upon the estate they have purchased to become intimate with the life and conditions of their surroundings. They may be lavish with their gifts, generous in their support of local institutions, but they do not really settle down, and always remain strangers. On the other hand some find their place almost at once ; strangers at first no doubt, their earnest desire is to become one of those among whom they dwell. Generous without prodigality, they seem to understand country ways and character by instinct, and bringing new ideas and business aptitude to a stagnant village they restore prosperity and self-reliance, finding popularity without hunting it. Mutual acquaintance ripens into friendship, and they take their rightful position and carry on the old traditions, sometimes better than those to the manner born.

Country society does not consist only of what may be broadly termed the landed gentry and the wealthy settlers, for it shades away gradually downwards to the local business man and the farmer, although every neighbourhood does not possess representatives of each grade. The retired Service man may be found almost everywhere, often filling some useful office. But the State is not a generous employer and it has till now usually rewarded men who have grown grey in its service, or been compelled by impaired health to retire, with a pension that is a pittance compared with their status in society. So they often retire to end their days in penurious independence if they have little means of their own, choosing districts where living is cheap and congregating in small towns where educational facilities are good. People who do not understand speak slightingly of snobbishness and exclusiveness, but, though the gibe may be occasionally deserved, this striving of cultured people to live on small means and educate sons and daughters in a fitting manner is one of the tragedies of country life.

Among the squire's friends is the doctor, one of the most hard-working individuals of the community, about at all hours of the day and night ; if you meet a motor-cyclist in the steepest and stoniest of lanes, or a car splashed with mud speeding along the highway in weather unfit for a dog to be abroad, it is probably the local doctor, for the motor-cycle and car have supplanted the horse and trap. The lawyers, other professional men, and the small manufacturers are found mostly in the little towns and large villages ; seldom landowners in the usual sense of that term, most possess a few acres and are prosperous enough. Their businesses have often been handed down from father to son for generations, and sooner or later most produce a shrewd member who expands his business and increases his landed property until perhaps he becomes the largest landowner in the neighbourhood, but against these must be weighed those who by lack of enterprise or sloth lose most of what they have, and the unfortunates who struggle in vain against the fierce competition of larger and better equipped rivals.

Comparative new-comers are the increasing numbers of middle-class folk who live in the country because they prefer its open air life, and those others compelled by ill-health to give up business before they have saved sufficient to provide a tolerable income. They live on the fringe of country society, and, unless they have resources in themselves and are content with their lot, their life is sometimes hard if no one of like station and class lives near. Their garden and their poultry yard provide occupation, and in the end they often become producers in a small way or earn fame for their strain of pedigree fowls, their pigs, or what not.

Many look on the supersession of the old landed proprietors with satisfaction and consider the wholesale breaking up of estates an advantage to the country, but this is not often endorsed by those acquainted with the facts and possessing some knowledge of rural England. In the past estates have constantly had new owners, and the development of rural districts near the cities is only the modern equivalent of an old story, but never have changes been quite so sweeping, and until the middle of the last century the town did not impose its rule on the country as it does now. The substitution of a soulless syndicate for the noble will be no advantage, and many small proprietors do not necessarily bring greater benefits than a few large ones. The mortgaged holding confers an advantage on no one, and the ill-equipped cultivator does not increase production. The War has taken a heavy toll of rural leaders and now more are going, for crushing taxation and high cost of maintenance, together with the set-back agriculture has received after the artificial stimulation of war, are making it difficult, if not impossible, to keep up an estate properly.   In consequence many which have belonged to the same family for generations are being sold, often the house and demesne finding neither purchaser nor tenant, and is left closed or in the hands of a caretaker, to the detriment of the neighbourhood.   If an impoverished landlord holds on to his estate for sentimental reasons it is usually as bad for the parish, as he dispenses with as many of his staff as he conveniently can, and responds grudgingly to his tenants' request for necessary repairs, while he has no money for improvements.   It will be an evil day for England if our landowners disappear ; for centuries they have had their place in country economy, a place they have filled on the whole to the advantage of all, not always be it said to their own.   True, many have failed to realize their responsibilities, but the majority have worthily done their duty to the nation and to their dependants.

MONTACUTE GARDEN FROM THE TERRACE, SOMERSET.

"THE GIANT'S HOUSE"—BREDE PLACE, SUSSEX.

THE PRESS.

THE MILL.

THE OLD-FASHIONED CIDER MILL—A HEREFORDSHIRE EXAMPLE.

OLD SUSSEX—A TEAM OF OXEN ON THE SOUTH DOWNS.

A FARMHOUSE ON THE YORKSHIRE HILLS.

THE DAIRYMAID—A DISAPPEARING TYPE OF "BEST DORSET."

SHEEP-DIPPING, YORKSHIRE.

III

FARMERS, OLD AND NEW.

Farmers, men of free nature and good condition ; some . .
have inheritance of their own ; others hold land by lease for a
rent . . . These albeit they be not so well accounted of (because
every man is of an aspiring mind), in times more ancient would
not have altered their conditions for any other vain-glorious
titles ; of whom many possess fair estates, keep up good hospitality,
and afford the very stranger hearty welcome . . . His chief
travails be most in matters of husbandry, wherein he leaveth no
pains untaken, whether it be by grazing, buying and selling of
cattle, or tillage.—*Risdon's " Survey of Devon."*

THREE hundred years old as this quotation may be, it is not an
inapt description of the modern farmer. Allowing for change of
circumstance and modern development the farmer remains pretty much
what he has always been. He gains more freedom in one direction and
loses it in another ; he and his labourers may be less dependent on
their lord but more dependent on the State. Science and research may
aid him in some degree, but it cannot altogether prevent him being the
sport of the weather. His conditions and work vary from district to
district and season to season in a manner incomprehensible to those
without practical experience. He has ever been raised and depressed,
waxing rich and adding to his holding when times and seasons are
favourable, facing ruin when circumstances over which he has no control
go against him.

If long settlement on the same property confer distinction not a
few farmers may scorn " any other vainglorious titles " ; thus, there
died a year or two back, a farmer whose family had owned and tilled
the same land at Maxey, Northants, since 1400, and longer connections
are not unknown. The " Statesmen " of the Lake District are also
examples of long ownership, but they are decreasing in number. What
were many of the younger sons of the great landed families but yeoman
farmers whose manor houses still dot the country-side to remind us that
agriculture is no mean occupation. Long tenancies are probably less
common than they were, but it is not very difficult to find holdings which
have been taken over by members of the same family in direct descent
for several generations, the tenancy, perhaps, showing greater continuity
than the ownership.

Long occupancy, however, cannot be deduced from the age of a farmstead, though its appearance and appurtenances may reveal something of the character and standing of owner or tenant. Farmhouses range from the one time manor house of a forgotten family to the humble homestead little more than a cottage ; of all ages and styles, built most commonly of local material according to local fashion, they reflect the methods and resources and climate of their district. From the warm brick and tile of Kent or the rose embowered cob of a sheltered combe to the deep porch and stark stone of the wind-swept hills, they are as diverse as the landscapes in which they are set. Usually picturesque in themselves or their surroundings, they are too often disfigured by patches of corrugated iron ; occasionally, alas, fire has wrought havoc with the old house, and there, surrounded by barns and stables and sheds whose lines and colours make a poem of architecture, flaunts a hideous villa of red brick. Sometimes, when the original house has disappeared, stable or tumble-down shed cannot quite disguise a chapel, and, rarely, a fine gateway or a range of wall showing signs of window tracery tells of a day when monasteries and their granges were familiar features of the country-side. Mayhap only the name of the original building remains, and although comparatively few can boast high lineage, the outbuildings and appointments of most, perhaps, testify to the usages of the past.

In the great cornlands are seen the huge barn and rickyards in which the corn was stored and stacked until threshed, and the still larger tithe barn has not disappeared. Outbuildings vary according to the size of the farm and its situation ; where the winter is long and cold, the cowsheds are larger and more substantial than in more favoured districts where the cattle are generally turned out for part of the day. The travelling threshing machine and the decay of tillage have made the large barn less necessary, so everywhere they are falling into ruin. The rick-settles of the stack yards have been used for fuel and the stone supports lie overturned amidst the weeds of summer. Even the cider cellar, that familiar feature of Somerset and Devon, is passing with the years, but in remote corners old Dobbin still plods round the horse-gear, cutting chaff and slicing roots. Where water power abounds a simple wheel is perhaps used for such work, but nowadays the harsh cough of the oil engine is often heard in the evening stillness, for the up-to-date farmer finds it a source of power that can be utilised at odd moments.

The farmer of old was not unduly bothered by science ; his crops were fewer and he followed the simple rotation of the period. His land regained its fertility by being allowed to lie fallow, and by the application of farmyard manure or lime. Stock raising, always of importance, especially the breeding of sheep for their wool, was not the scientific industry it has become during the past century, and animals were used for purposes which to us are unfamiliar. Oxen drew the plough, and cheese and even butter were largely made from ewe's milk. Sometimes cow's milk was added, for Tusser says :—

> Five ewes to a cow, make a proof by a score,
> Shall double thy dairy or trust me no more.

Camden records that in some parts of Essex the ewes were milked by men, and other writers mention the cheeses for which the Essex marsh farms were famous. Wool was always in demand, and when labour was scarce after the Black Death landowners found it more profitable to turn their tillage into pasture ; the development of weaving encouraged this to such a degree that Parliament at last intervened and determined the number of sheep a farmer might maintain, according to the size of his holding. At the beginning of the tenancy system the stock was usually leased with the land, a method that has not died out on the sheep runs of north-west England. Sheep may no longer provide milk nor oxen be kept for ploughing, but the important influence, direct and indirect, of livestock on agricultural economics is little understood by those unconnected with farming.

Manorial and legal records throw light on the agricultural life and methods of the fourteenth, fifteenth, and sixteenth centuries, and, incidentally, show that some of the problems of to-day are not so new as often supposed. Laws to prevent plough-land being turned into pasture, to fix wages, and to provide the cottager with land or allotments have their counterpart in days gone by. Wills give an idea of a farmer's possessions and the implements commonly used, also showing the close connection between the farmer and the merchant. In the sixteenth century a feather bed was sometimes a cherished heirloom, articles of pewter are often mentioned, and farmers of the better class usually had some silver. Implements were few and simple, and more than one cart is seldom mentioned, but, indeed, these were not unduly numerous in remote and hilly districts within the past one hundred and fifty years.

Towards the end of the sixteenth century and onwards the writers take up the tale and it is possible to reconstruct the rural life of many parts of England without much difficulty. Henry Best, a well-to-do farmer of the East Riding, has left his " Farm Book " to draw the twentieth century very close to the seventeenth in many ways. He bewails the high wages asked and the difficulty experienced in obtaining good dairymaids and servants, although twenty-four to twenty-eight shillings a year for the two latter does not seem exorbitant, the men being paid in proportion.   The hinds, a word still used in some districts ; were hired at Martinmas, the parish constable drawing up a list of masters requiring men, and men masters—a custom continued up to recent times.

Even a century later wages had not risen much, men and maid-servants being given an advance when they wanted to purchase anything or go a-fairing.   It was customary to allow so much beer or cider a day, and to this must be added a patch of potato ground, a pig or side of bacon maybe, and, probably, a certain amount of wheat, if not free at least at cost price.   The wage of the farm labourer has never been so low as it appears, as to such allowances, even now continued to some extent, must be added a cottage free or at very low rent, harvest money, and the extras earned by hedging and other odd jobs paid for at piece rates.   Poor Law children, both boys and girls, were apprenticed to farmers, a system which has been severely criticised, but the indentures were usually strict enough and if properly carried out by masters and those who looked after the children provided an excellent training. Even now a small boy from the Institution (workhouse), who makes himself generally useful and has every opportunity of learning farming, is not unknown.

Farmers are famous for their good living and though the fare was formerly more homely, not to say coarser, there was usually plenty of it.   The white loaf may have appeared in the smaller farmhouses only on festive occasions, and the labourers had to make the most of rye and barley bread, but it was seldom stinted, and cheese was plentiful, though for dinner the men might have to put up with dumplings boiled with the family's bacon.   Meals were—and are—provided for the workers at harvest, when they were more generous and tasty, and home brewed ale or cider could be drawn on to almost any extent, many a labourer still estimating his master by his generosity in this respect.   Master and men had meals together at one large trestle table in the kitchen, family at one end, labourers and servants at the other ; here and there the old patriarchal custom has not quite died out, or the men may come in when the family has finished.

Labour may have easier conditions and better pay, and science aid the farmer in many ways, but the latter is not a care free individual spending his time riding round his land with leisure to indulge in hunting

A CORNER OF AN OLD WELSH FARMHOUSE KITCHEN.

A TRAVELLING BAND OF SHEEP SHEARERS AT WORK.
"They sheared in the great barn."

A WELSH SMALL HOLDING.

AN OLD DEVON ORCHARD—UNSCIENTIFIC BUT FRUITFUL.

and shooting. His life is chiefly an effort to do two hours' work in one ; each day bringing its appointed tasks at the appointed time with a crop of problems to be solved and difficulties to be overcome. The cows must be milked at the regular hour and the milk measured out for the round, or if sent into the city the churns must be despatched to the minute. Milking is pleasant enough on a summer day but very different on a freezing winter's morn when the fingers are almost too numbed to hold the lantern whose flickerings hardly dispel the gloom. Lucky indeed is the individual who hears the words of an old milking song float down the lane, for, except on the smaller farms where this task falls to wife and daughter, the male milker is the rule, and the machine will doubtless oust him in time. In many areas the dairy is becoming of minor-importance, for the " milk factory " sends forth its motor lorries to pick up the churns, the milk being sent away or made into butter, cheese, and milk powder. Where these factories exist, churns will be seen at the ends of lanes awaiting collection or standing on roughly constructed wooden platforms to facilitate loading.

Maybe a sick beast requires early attention, and after a feed the horses are harnessed for cart or plough, after which the yard may be quiet for an hour or two. But except perhaps during haying and harvest some work is going on round the farm. Implements require overhauling, shippens must be cleaned out, chaff cut, water pumped, stock fed or brought in from the fields, manure carted, and so on according to the season. Perhaps a prize animal is being prepared for exhibition, requiring special feeding, grooming and exercise, or there is work in garden or orchard.

Sheep shearing means a day or two of continuous work ; if the flock be small the farmer and his regular hands require no assistance. It may chance that a large barn is the place of shearing, the sheep coming in at one door, and, naked and frightened, passing out at another. A meadow near a stream may be the scene of operations so that the washing can easily be done before shearing is started. Deft and quick are the shearers, turning the struggling sheep this way and that until with a final snip the work is done, the fleece quickly rolled up and tossed aside. A pot of tar stands handy to dab on any wound, but the regular shearers are experienced and seldom hurt the sheep or spoil the wool. Bands of travelling shearers are less common than they were, and the clipping machine driven by a petrol engine is supplanting the hand worker where flocks are large. Altogether shearing is not the festival it used to be when the farmer invited his friends, and shearers and shepherds, with fun and feasting, elected their king.

Threshing time was formerly a busy period in the yard, but now the grain is mostly stacked in a convenient field near a gate ready for the travelling threshing machine, although the rickyard with its golden stacks is still common in some parts. The horse thresher is somewhat rare and the flail has become a curiosity, the threshing floor in the barn alone remaining to tell of former days. The flail was a cunning instrument needing cunning use, and its component parts were known by different names in different parts of the country. Perhaps made by its owner with some help from the blacksmith, the " frail " of Sussex and " dreshel " of the West Country was often one of the simple yet useful heirlooms of the labourer. Threshing was followed by winnowing, a slow process when the doors of the barn were opened and the wind did the work or when the corn was shaken out of a sieve before a fan. Next came the winnowing machine worked by hand, often to be found among other out of date implements in the " museum " corner of a shed, but now threshing and winnowing are carried out as one operation.

An oast house is the sign manual of Kent as the cider cellar is characteristic of Hereford and the west. A group of red roofed drying kilns, topped by the hoods whose outstretched vanes terminate in a hand or cock or what not, is more picturesque than the cider cellar. After the hops are weighed in the field where they have been gathered they are carted to the kiln for drying, which is a more scientific process than of yore, automatic devices giving an alarm if the temperature become too hot. When dry the hops are packed tightly into sacks called pockets to be sold. Cider making is a more lengthy operation, for after the apples have been knocked from the trees by poles they are gathered into heaps for a few days before being taken to the mill. Bruised in the mill they are built up on the press into layers between clean straw, the juice running into stone troughs or wooden tubs as the pressure is increased. The old cider mill, not quite extinct but seldom used, was a circular trough in which ran a wheel of the same material. Formerly sulphur was burned in the half-filled casks to make the cider sweet, but now this is achieved by racking or pouring it from one barrel to another.

Endless are the pre-occupations of the farmer. His horses must be shod, his waggons require the periodical attention of the wheelwright. Sheep must be dipped at the appointed time, the visit of travelling stallion or pedigree bull arranged, and a certificate obtained from the police for the removal of the carcass of a cow that has died, and perhaps the breaking in of a colt superintended. Wandering sheep or bullocks

A NORFOLK FARMSTEAD.

THRESHING TIME—NEW STYLE.

AT THE DOOR OF THE DAIRY.

"The farmer's a man of brawn and thew,
With cheeks and nose of a ruddy hue."

must be brought back and, in a dry summer, visits to far off watering places may be necessary. Unexpected visitors drop in at all hours, but the evening is the best time to find the farmer at home though he may be inspecting stock at distant pasture or selecting animals and produce for market.

The farmer's wife has plenty to do, for in the first place the household management of a farm is no light task. The dairy is her province, the poultry her care, and preserving and pickling her pride. The cheese press is seldom in evidence, though even in parts where little is made some good wife may pride herself on her cheese, but butter-making is important, while it is her business to look after the local milk round, with or without an assistant. The old fashioned dame of Devon and Cornwall still beats the cream with her hand, disdaining the churn. In the same districts, too, the making of scald cream is an art in which she delights ; before the advent of the separator the milk was set aside in wide shallow pans for the cream to rise, when they were placed over a fire, of wood for preference, Now the pans of separated cream are ready almost at once and put on to the stove until a crust is formed when they are withdrawn and cooled quickly. The separator yields more cream but must be kept scrupulously clean so hardly lightens labour in the small dairy.

In spring, when hens are going broody and chicks and ducklings appear, she is busy with her poultry, and, perhaps, wrapped in a blanket in a box beside the kitchen fire a brood of young ducks begin their acquaintance with the world. Every evening she collects her birds and sees that each mother has her proper brood before shutting them away from marauding rats. For the rest of the year the poultry do not trouble her unduly unless she has no children to collect the eggs. At Michaelmas, Christmas, and seasons when consignments of poultry are sent to market, plucking parties were once common, neighbours and friends being invited to help before sitting down to a substantial supper. Hers, too, is the task of preparing the dairy produce she takes in to market in baskets covered with spotless cloths and bowls that glint in the sun.

It is not easy to draw a picture of the typical farmer as there are so many of them. At one end of the scale is the large landowner with a model farm who has done so much for cultivation and stock breeding ; at the other the holder of a cottage, a shed, and a few acres, whose garden, pigs, and poultry provide most of his income, though he may buy young calves or pigs to keep a week or two before selling again. The holders of from one to fifty acres form the most numerous class, but perhaps the most typical farmer is tenant or owner of some 200 acres, a man who has to work hard all the year with scant opportunities for leisure, as he must always be ready to make the most of the weather and the best of his opportunities.　He may not trouble himself unduly about accounts, but farming does not lend itself to easy book-keeping, and his rough annual balance sheet tells him whether he is solvent.　Often a shrewd man of business in spite of apparent carelessness, he takes some beating at a bargain ; he is always " the master " though his wife is sometimes " the better man," and the widow who farms successfully with the aid of sons is known everywhere.

The merry-makings of harvest and the rubber of whist with cronies at the local inn belong to another day, but where the weekly market is still an institution he would on no account fail to don market clothes and sally forth on horse or in trap to combine business with pleasure. He seldom walks if he can ride, and though he rarely has time to follow hounds he will be met in the fields as dusk is falling, gun on shoulder, dog at heels.　In the spring he entertains a party of friends to thin the rooks and wood pigeons that take toll of his crops, and no one enjoys the rough and tumble sport of the harvest field more than he.　Learned he seldom is, and his reading is not deep, being restricted to the news-paper and a farming journal ; if either be neglected it is the former. Being the typical Englishman he grumbles, for he has more cause to grumble than most men.　A long spell of sunshine at the wrong time may do more harm than prolonged foul weather at a season when it is expected, but both cause bursts of feverish activity to overtake arrears.　So he grumbles and goes about his business, setting a good example to most of his critics.

When all is said and done the English farmer follows his calling as much for the love of his craft as the profits it brings.　His occupation with its ever varying work, its small but absorbing interests, the changes of the seasons, and converse with men of like habit to himself, satisfies him as would no other.　He pits his skill against the changes of the weather and gambles with circumstances over which he has no control. If he were not an optimist, if he thought only of profit, the lean years of the past century would have seen his extinction, but he takes the rough with the smooth, apparently seldom satisfied but really more content with his lot than most men—provided always there are not undue restrictions on his freedom of action.

A THRESHING FLOOR—OLD STYLE.

"Thump after thump resounds the constant flail,
That seems to swing uncertain, and yet falls
Full on the destined ear."

THE OLD BARN.

. . . . " I hear the rickyard fill
With gossip as in generations gone,
While wagon follows wagon from the hill "

WEIGHING PEAS.

A SPRING TASK—ROLLING WHEAT.

IN THE MODERN HARVEST-FIELD.

THE MIDDAY MEAL.

IV.

FIELD WORK.

ALL the work of the year on hill-side and in valley is summed up in the expression Seed Time and Harvest, which represents such a variety of crops and occupations that the mind is bewildered. Almost always some seed bed is being prepared, some harvest garnered, and as the seasons move on their ordered course they take their tasks with them. One county has started to plough the stubble ere another has begun to cut the corn, and if you survey a landscape which includes hill and dale you may see both in progress. Not only are there crops common to all England, although they flourish most in special parts, but many so small and local they are apt to be overlooked.

Methods of cultivation, though broadly similar, vary in detail according to local experience, climate, and soil, modified to some extent by the class of farming adopted, the requirements of the large dairy farm and the mixed holding differing considerably. Implements, too, are astonishingly diverse ; the most up-to-date machine may be of small utility in one area or an unkind season bring back the methods of our forefathers. In the hand-tilled fields fashions in spades and mattocks are noticed, together with special tools bearing local names seldom used elsewhere. In remote districts implements of all ages may be seen in yards and sheds, rusting beside the smithy, or perhaps buried in the wild-growth by the gate, whence they are occasionally hauled forth to labour.

From time immemorial agriculture has developed slowly with bursts of rapid improvement ; the outbreaks of the Black Death left their mark on cultivation, but progress hardly began before the end of the sixteenth century. Illuminated MSS. and woodcuts in the earliest printed books show that although operations may be largely the same, methods have vastly improved. With the enclosures and the intro-duction of new crops from abroad our landscapes began to resemble those of to-day. When iron became plentiful, agriculture benefited in many ways, and the wooden plough was shod with metal, a form that lingered in odd nooks up to quite recent times. The efficient reaper and binder is modern but the drill made its appearance some two hundred years ago, a Scotsman being the pioneer.

55

Many regret the general introduction of machinery, declaring that all the poetry has gone from field operations. Certainly the tractor plough is not picturesque, especially one of those bluff nosed models which resembles some uncouth monster as it heaves into sight out of a hollow, enveloped in a cloud of vapour and shaking with stertorous breathings, but it is better than the steam plough pushing its many furrowed way across a large field like a huge spider. It will be a sad day when the teuf-teuf of the motor replaces the jangle of harness and the pleasant hiss of the plough cutting the furrow, and the reek of petrol is but a sorry substitute for the scent of newly turned earth. No doubt the hayfield lost something of its beauty when the line of men advancing with scythes flashing in the sunlight was supplanted by the mowing machine, but few can maintain that the progress of the harvesting machine, drawn by its team with tossing heads, throwing aside sheaf after sheaf in regular line, is a less pleasing sight than a group of men laboriously reaping with a sickle.

However, it will be long ere the team is banished from the field, and those with time to idle may still watch its varied work through the year, beginning in autumn when the ploughman, followed by a crowd of jostling rooks, drives his regular furrows from hedge to hedge. Where fields run small and are irregular in shape, steep, or dotted with boulders, the tractor will not quickly oust the horse, so the old-fashioned ploughman still holds sway, and only a few more varieties are added to the ploughs in use, for almost every farmer has his favourite and every soil the model that suits it best. In Feudal days the ploughman was an important individual who possessed certain rights ; he received so much land or corn and could use the lord's plough, but his duties were strictly defined and he was commonly required to plough an acre a day. The oxen were also under his care ; eight usually went to the team, and they were driven much as they are now in the few places where they are yet in use. About a century ago they were comparatively common in Wales, the North, Devon, and the South Downs, but the few teams remaining in Sussex and near Cirencester are regarded as curiosities.

Where the soil is light, the plough may have a rest for a season, the cultivator preparing the ground sufficiently for certain crops, but if the land be heavy, repeated harrowings must follow the first breaking of the stubble, for ploughing is only the beginning of cultivation. What more pleasing sight on a sunny day in autumn than a large field being sown ? At intervals along one margin are the sacks of seed from which

REPAIRING THE WIRES IN A HOP GARDEN
OF KENT.

THE APPROACH OF HAYSEL—SHARPENING THE
KNIFE OF THE MOWER.

PICKING TIME IN A KENTISH HOP GARDEN.

HAY TIME.
"Men that load the sluggish wain."

the hopper of the drill is replenished ; in front goes a harrow, not far behind follows the drill, then another harrow, with perhaps a roller bringing up the rear. What a bridging of the years sometimes occurs when these operations are carried out together ; the drill may be fresh from the factory, bright with new paint, the harrows of different models and ages, and the roller of stone, still commonly used where it can easily be obtained. Here and there if the soil be light a still older one is pressed into use when the season lags—of wood bound with iron.

Broadcast sowing, an operation requiring more skill than is apparent to the onlooker, still has its uses. In very small fields it serves its purpose well enough, and sometimes in spring the broadcast sower, his shallow basket or tray held before him, goes forth ; with eyes fixed on a mark at the opposite hedge to keep his line, he walks to and fro, the sweep of his arm keeping time with his footsteps, sowing dredge corn, usually oats, often done if there be a fear the first crop may fail. Spreading manure is a prosaic operation essential on our land which has been tilled for centuries. Once the spreading of farmyard " muck," it is now a work of science, for your up-to-date farmer weighs the relative merits of artificial fertilisers, and the truly scientific cultivator calls in the chemist to tell him in what his land is deficient. Along the sea borders evil smelling seaweed or fish, which offends the nose for miles, helps to restore fertility, but lime has had its day, to which fact the crumbling kilns bear mute witness.

Spring is an even busier season than autumn, for the last fields must be ploughed and sown, and sprouting winter wheat rolled. Grass-land, too, requires its meed of attention, and as winter turns to spring the farmer is busy in the meadows, carting and spreading manure, attending to ditches and drains. If rushes grow they must be mown regularly, and when the first warm airs of spring herald the awakening of nature the roller and chain harrow are brought out to level the ground and complete the spreading of manure, in the process painting broad bands of light and shade across the meads. Artificial fertilisers may also be applied by a machine resembling a drill.

With the dawning of spring, too, come the auctions of grazing ground and meadow, and when the delicate hues begin to turn to less variegated green the migratory labourer appears on the scene. Less common than formerly he still finds plenty of work, for machinery and rural depopulation have reduced the reserve of local labour. The Irish labourer was more common in the north than the south, but in most parts the migrant could be encountered between haytime and potato lifting. He camped beside the hedges of Middlesex, and when the hay was in, he passed from farm to farm in the corn growing districts. In some parts of Surrey whole families would combine business with pleasure by taking part in the harvests of Sussex, but they are seldom seen now and the motor has dealt a severe blow to the hayfields of the counties adjoining London.

As the time for mowing arrives the farmer looks anxiously at the weather, the blade of the mower is sharpened, the tedder and horse rake tinkered and oiled, and waggons sent to the wheelwright for a hasty overhaul. The beginning of haysel is still a great day, but the ring of whetstone on blade is no longer the orchestra that plays the overture to the proceedings. However, it remains a village festival, and where labour is scarce and small farms abound all troop into the meadows, leaving the street empty and deserted. In the late afternoon the farmer's wife drives out with her children to take tea with the workers in the shade of a hedge, and if the farmer be not directing operations he comes out to see how work is progressing.

This is the beginning of the real jollity of the day, as the mowing is all but finished and the time has come to turn the swathes. The children romp and there is constant laughter as the men and girls chaff or make " sweet hay " with a new-comer. Carting is eagerly anticipated by the boys who aspire to lead the horse. Then away to the yard or a corner of the big meadow where a layer of brushwood forms the foundation of the stack. As it rises higher the elevator is brought into use, the builders laying the hay carefully and evenly, as if improperly built or put up wet, it may overheat and catch fire. When the stack has settled down the thatcher arrives, a solitary worker who finishes in silence the task begun and continued in cheerful noisy concert.

Haymaking takes off the edge of enjoyment, and though the harvest field may not lack frolicking it is no place for the idler, especially in uncertain seasons when its work must be completed while the weather allows. The harvester regularly tossing aside the finished sheaf is the epitome of efficient machinery, and there is an appearance of half-used strength about the arms of the machine which only cuts and throws out sufficient loose corn for a sheaf. Behind it a little group moves slowly along the rows binding the sheaves, and in their rear others build the shocks, setting to partners and turning away. A stook must be set up to allow the breeze to blow through but only a hurricane over- turn, and barley containing more moisture is built of fewer sheaves. Perhaps in a corner, two or three labourers cut with a scythe the corn

OAST HOUSES WITH THEIR MELLOW BRICK WALLS, RED ROOFS, AND WHITE COWLS—
IN A WORD, KENT.

TEAMS AFIELD.

BUILDING THE ROOT CLAMP

badly laid by a hail storm, and occasionally the sickle may be used. In the fields of a little farm a single labourer may help the farmer cut, binding and shocking being done by wife and family.

Meals in the harvest field are merry enough, and the farmer is generous with his fare and ale or cider ; if he were not, his labourers would ask the reason why, as they have become fastidious and occasionally grumble at the wholesome food provided. Hereford, Somerset, and Devon are the cider counties, four quarts a day being the old allowance, but the copious libations of cold tea now so universal are not appreciated so much. Fast and furious waxes the fun when the last square is being cut and the rabbits bolt in all directions. What a shouting and barking of dogs for a few exciting moments as the quarry is killed with shot or stick by the farmer and his friends and men, who lay their bag under the hedge beside their coats. Clearing the field has its enjoyments, and once again the boys compete for the honour of leading the horse, or, better still, driving the waggon to rick yard or corner by the gate on the lane, handy for the travelling thresher.

Neither haysel nor harvest is always merry, for in damp seasons the sweating horses can hardly pull the machine, and the hay must be made and re-made until it loses all virtue for feed, or the harvester put aside for the older machine that cuts without binding, increasing labour and cost.

Much might be added of harvest methods in different counties, of steep fields where the binder cannot cut the fourth side of the square, of the variety of waggon, from the great, bluff, three or four horsed wain of the plains to the two-wheeled cart and primitive sleigh of the hills. The end of harvest is not the festival it was ; the last ears are no longer twisted into the Kern Doll and escorted to the farmhouse with proper ceremonies, and the harvest supper hardly obtains. Gleaning has gone, and where the gleaning bell survives it has lost all significance.

Cereals are the best known of our field crops, but clover, sanfoin, rotation grasses, together with roots, potatoes, and others variegate the carpet of our landscapes, in some largely entering into the main design, in others only serving to emphasize the predominance of corn. The townsman hardly realises the influence of live stock on cultivation. On most farms the animals determine the rotation, for fodder crops, both green and roots, are as important as cereals. Clover sprouting green amidst the stubble is a familar sight ; sheep folded on roots not only produce mutton and wool but manure the ground for the following crop ; if climate and soil be favourable catch crops for feed allow the fullest use to be made of the land.   This explains the reluctance of farmers to put down an undue proportion of wheat or break good pasture, for if the balance of cultivation be upset profits may be reduced and less rather than more food produced.

Thus the farmer draws up no hap-hazard programme for the year, but has to consider carefully a complicated problem, for the rotation depends not only on local conditions but is modified to some extent by circumstances.   Prolonged wet may ruin the hay and play havoc with the corn, causing a late harvest ; fields infested with pests or over-grown with weeds require a cleaning crop, which needs nice adjustment to fit in with the others.   Experimental farms are endeavouring to produce plants able to withstand bad weather and attacks from insect pests and also yield more heavily.

Compulsory education has killed the bird scarer, the boy who sat on a gate or hid in a hedge twirling a rattle or firing shots, but birds have not mended their ways and are reinforced by feathered aliens which have discovered England to be a land of plenty.   When wind and rain lay the corn birds can pick out the grain at their leisure ; if harvest be delayed every ripple of wind begins threshing before its time ; standing long in stook causes it to sprout ; and even when safely stacked rats take some toll.

Kent is the home of all the farming arts and one longs for its diversity when passing through country in which pasture or the more monotonous crops predominate.   Essex, Suffolk, and Lincoln are, perhaps, the counties for peas and beans, and Norfolk for roots.   A devasted pea-field is not a pleasant sight, nor beans shrivelled almost black, but well tilled roots look trim and tidy.   In late autumn the building of the clamp in a sheltered spot by the hedge may be watched, and by an odd root dropped here and there by the wayside the sheepfold discovered.   Here is the slicing machine, and for a change of diet coarse cabbages are grown. Like roots these require much weeding, and if labour be short or the season wet the rows are lost amidst a veritable jungle until the horse hoe or line of labourers restores tidiness.

A row of hoers bending over their task when the baked earth sets the air a-quiver and the animals seek the shade, presents the very essence of monotonous toil, and no one can regret that the woman worker becomes more rare.   Except in some districts and some occupations the woman field-hand is disappearing, although the war brought her back to some extent.   In Northumberland she is not uncommon, especially for turnip hoeing, and is called by the old name of Bondager,

POTATO DIGGING, GREAT BUDWORTH.

GATHERING WATERCRESS AT EWELME.

which has lost its original meaning. In the past, labourers in the farm cottages had to provide an assistant, usually a woman ; gangs of them, wearing the big straw hat and kerchief, the traditional costume, still labour in the fields, but the system itself has vanished.

Every farm has its potato field, but in Lancashire, Lincoln, and Cambridge, potato growing is a great industry in which machinery and organization are all important. In sunny pockets along the cliffs of the south-west little patches facing south produce early potatoes ; bordering the Ouse in the East Riding are fields notable for second earlies, while in the big main-crop districts the tubers are harvested by machinery, which lifts, grades, and bags. With the potato may be coupled the market garden, universal on the outskirts of great cities ; dreary as are these expanses of vegetables diversified by bush fruit, with glass houses at intervals, they are an important branch of agriculture.

The cultivation of hops is enough to whiten the hair of the most hardened farmer. When he has weathered a prolonged winter and months of careful tilling promise a bountiful return, some insect pest proves his undoing or a violent storm turns his ordered garden into a tangle of wire and poles. Such visitations may be absent and all is ready for the picking when continuous rain dashes hopes. Formerly numerous farms grew a few hops to brew their own beer, and hop gardens on a larger scale are not uncommon in several counties, but now Kent remains supreme, with Hereford, Worcester, Sussex and Hants far in the rear. Except at the end of picking when the gardens are a litter of rotting bine and debris, a hop garden is always attractive, whether the old style arcades of poles or the more modern walls of wire strands, the elaborate system being erected by men on stilts. Early in the year such a garden is curious rather than picturesque, but in summer the wires are lost amidst the bine, and the sprayers at work spread misty clouds of vapour that veil the middle distance.

Picking becomes more orderly and methodical every year ; the tally man with his wooden tablets has given place to the clerk with his ledger, and shelters erected under the eye of a sanitary inspector provide greater comfort than the makeshift shanties and tents that were the rule. Hop pickers are not invariably the poorest of the cities seeking change in the open air, as some places prefer local labour. When dusk has fallen and the evening mists begin to rise the twinkling fires breaking out in lines and groups turn a Kentish valley or a Hampshire hill-side into fairyland ; the smoke billows upwards, and the children pass to and fro in the flicker like elves in a garden lit by glow-worms. That is the pleasant side of hopping, but in seasons when cold rain churns the ground into quagmires, when wet enters the driest shelter, it is another matter, especially if shortage of hops means fruitless journeys for pickers, who drenched and penniless, drag weary limbs to the nearest workhouse.

The hop grower may bewail his lot but his neighbour with extensive strawberry fields faces great risks with uncertain profits. The Garden of England still stands at the head of strawberry counties, but Cambridge and Norfolk now grow as many as Hampshire. From the Mendips come some of the earliest, and in East Cornwall are sheltered valleys where the odour of strawberries lies heavy on the air and school holidays arranged to allow the children to take part in the pickings. Strawberry pickers belong to a migratory class from town and country, some seeking a profitable change, some taking part in the less skilled occupations of the seasons.

It would be wearisome to mention the many crops, universal and local, the cultivation of which makes up the sum of field work. Some show how closely agriculture and industry are related, others can hardly be said to lie in the domain of work in the fields. To the former belong mustard, which gives colour to so many acres in Norfolk and Cambridge, and flax, not beloved of farmers as it takes so much out of the ground and needs a deal of weeding. Until brought back by circumstances it was remembered chiefly by field names, by abandoned flax mills in the districts where it was common, and in Dorset by the" lynchets " or hill-side terraces on which it is said to have been dried. Teasels were also extensively grown in Somerset for the textile industry and as machinery has not supplanted them for certain cloths a few acres near Taunton are devoted to them. Woad can seldom have been a large crop and the famous farm and factory at Parson's Drove, near Wisbech, has been closed. To the factory also, for the distillation of its oil, goes peppermint, a small and local crop requiring the right soil and expensive cultivation, and now that jam making is such a large industry it is migrating to the country to be adjacent to the fruit garden.

Bedford is noted for Brussels sprouts, the country round Leeds for rhubarb, and Evesham not only for fruit but asparagus, a day's cutting of which may represent twenty tons. The trim orchards of Kent are a great contrast to those of Devon and Somerset where the apple trees prove that if conditions be favourable they can flout the rules of good horticulture. Bucks and Surrey, with other parts, find no little profit in watercress, a field crop or waterside industry as you please. Scilly is the home of early blooms, and the small flower farm flourishes in other favoured localities. Somewhat similar is the raising of seeds, for British seeds are famous the world over. Railway travellers are familiar with some of these centres, for the chequers of colour are bound to catch the eye, and round Kelvedon and the Coggeshalls in Essex an extensive area is devoted to little else, introduced so it is said at Little Coggeshall by the Cistercians.

This passing glance at the varied work of the fields, familiar and unfamiliar, is sufficient to show how great and all-embracing is the calling of the farmer. Agriculture changes but never stands still, once important crops may disappear but others take their place. Only by unending toil and never ceasing attention are they brought to fruition, and when labour and experience have done their best circumstances beyond control may rob the farmer of his hardly earned profits. An over-bounteous season may prove nearly as ruinous as an unkindly one, so we should treat his grumbles with lenience, and even the sailor turned landsman would find another side to his inevitable " Who would sell a farm and go to sea ! "

V

COTTAGE FOLK.

ENGLAND alone among the nations of age long settlement, perhaps, possesses no real peasant class. Thousands of families may for generations have seldom been more than humble toilers, but they have never been a race apart, completely cut off from those above them. Thanks to our chequered and adventurous history—invasion, civil war, plague, expeditions overseas—extravagance, plodding industry, and the claims of talent, the nation has always been on the boil. Seldom has it been impossible for the lowly to better themselves, though opportunities may not have been so many as to-day, and never have the more fortunately situated ceased to sink in the scale for one reason or another, each class thus leavening the others.

Thus it is that the servile peasant is not typical of rural England, though it were fatal to generalise, as in one district the cottager is an independent man of open bearing, giving deference to his betters without either servility or truculence, and in another he obviously feels his inferiority, showing it in manner and speech. He is as diverse as his environment as scores of folk rhymes and jibes testify, but ease of communication and a dead level of education are destroying many local peculiarities of character and speech that once were so marked.

Old writers frequently allude to these characteristics, some of them rather difficult to believe. Dr. John Burton, touring in 1771, asks somewhat ungallantly in his " Journal," " Why is it that the oxen, the swine, the women, and all other animals, are so long legged in Sussex ? May it be from the difficulty of pulling the feet out of so much mud by the strength of the ankle that the muscles get stretched as it were, and the bones lengthened ? " Sussex remains muddy to this day, and more than one writer has coupled the mud of the county with the slow clod-hopping ways of its rustics, while the Cornishman has long had the reputation of being a gentleman.

It is not easy to give a well balanced picture of the English cottager, which is the reason he is usually described as a wretched down-trodden slave, living in an insanitary hovel, or as a light-hearted peasant dwelling in a rose-embowered cot in Arcady. Especially difficult is it

GOSSIP.

"They chat of various things,—
Of taxes, ministers, and kings ;
Or else tell all the village news."

THE FARMER'S BOY.

A FAST-VANISHING TYPE.

at the moment, for the war has caused an upheaval that will take years
to subside. He—and she—is so elusive and varied that he is nearly
always represented through the spectacles of the observer; we get
either a passing snap-shot or a local portait, miniature like in its details.
The latter is usually the more accurate, but it may be quite false if taken
to represent those who dwell on the other side of the hill. The cottager
can only be judged in relation to the conditions of his district and the
circumstances of those who live around him. What may be abject
poverty in one place is comparative affluence in another, and least of
all can his prosperity be estimated on a city basis. Pitifully small his
income may often appear—at least before the war—when reckoned in
money alone, but in the past the country has never been run on a strictly
cash system.

Here and there are grasping farmers and fearful labourers, but they
are fewer than is generally supposed. Skilled labour is seldom so
plentiful that the bad employer is unduly favoured, and he who has
unpopular views on beer or cider is not always able to pick and choose
during periods of rush. The labourer knows the reputation of every
farmer in his district, for their delinquencies are discussed at the inn.
Whatever the future may have in store the weekly wage by no means
represents the cottagers' full earnings at present. He is paid piece
rates for extra tasks, and to this must be added harvest money, and the
supplies which vary from district to district. He seldom pays an
economic rent, he has his garden or allotment—usually selling his surplus
produce— and something may be made out of the odd jobs always
going in the country.

Some cottagers keep poultry or bees, a pig or a few geese if they
have room, and no little money is earned by the children according to
the season and the locality. Daffodils and primroses, whortleberries
and blackberries, sloes, mushrooms, nuts and acorns, all bring in a little,
though it is a fortunate neighbourhood that has them all. The cottager's
wife went charing or took in washing, but high wages are making her
independent and such occupations are now considered derogatory. A
very few may be clever needlewomen or noted for a special dish or
preserve which helps to add to the family income.

Although conditions vary so greatly that it is unwise to generalise,
it may be said that on the whole the cottager lives very well. Good meat
is within the means of all, and during the war frozen mutton and mar-
garine raised a louder outcry in some parts than greater hardships.
There is an increasing demand for tinned food and kickshaws that are

easily prepared, but do not contain such solid nourishment as the former pork and cheese, and it must be sadly admitted that twentieth century economy is often very wasteful.   In the back-ways the wife who bakes her own bread in the simple oven beside the hearth may occasionally be encountered, and very proud is she of the fact.   Some, not possessing the usual oven, baked the bread and cooked the meat in an iron pot amidst the embers until the village baker undertook it for a trifle.   In the open fire era, the pot was more in evidence than the pan, probably of more benefit to the family with its stews and hotpots and substantial puddings than the more tasty and usually more greasy " fry " of to-day. The open hearth possessed a battery of pots and pans, with their hooks, hangers, trivets, tilters, and so on, simple utensils, artistic and lasting, yet easily replaced by the village blacksmith.

As the cottager's food, with the exception perhaps of fat bacon, approximates more and more to that of the townsman so do his clothes. Long ago he ceased to wear home-spun and the smock, which have given place to the cheap slop suit of the market town, an elegant affair indeed when new but without a quarter of the wear of the old garments. You must walk many miles ere you happen upon the old " round frock " that outlasted a lifetime if made of good linen.   Now, the death of one who clung to the old fashion is chronicled in the newspapers, yet some such figure can be recalled, an aged Sussex farmhand whose work it was to attend the prize bull, a small farmer of Hertfordshire who to the last wore his green smock, the colour of the county.

A wealth of romance and ancient lore has been lost with the smock. Sweethearts embroidered them for the Sunday wear of their future husbands ;   the cut and colour and decorations varied with the district and the calling, and if worn by milkmaids were usually worked with hearts.   Their use was not universal and authorities differ as to the extent of territory in which they were commonly worn.   The ancient " topper " of beaver appears occasionally at a village funeral, the huge hat of coarse straw is to be seen in the harvest field, and the soft black hat of the traditional rustic has not lost favour with some of the older men, but the young ones prefer the cap for work and the soft felt of Continental shape for leisure.

Few of the girls are content with the simple cotton frock as they desire something more " fashionable."   If well brought up they spend much time making what they fondly imagine is the latest thing, and if not they go to the village dressmaker, who always finds plenty of work, for the country woman of to-day is seldom clever with her needle.   However, she is skilful in adopting clothes for the small boys, and round the coast everlastingly plies her knitting needles, making jerseys for her men-folk, old and young, or in some districts, socks and stockings.

Costume no longer bears the stamp of district ; it would be hard to find a Welsh woman in high hat, for instance, but certain broad distinctions may be noticed. The shawl and the kerchief belong to the north rather than to the south, and the overall-pinafore is almost universal in the south-west, while the clean white apron is indispensable to the market woman. The frilled cap worn inside a bonnet is still sometimes favoured by the oldest women ; the sun-bonnet is going, though the girls find it not unbecoming for the hayfield. Cottagers spend far more on dress than formerly, but it cannot always be said that the result is either suitable or becoming.

Perhaps, one might hazard an opinion that the cottage woman is more house-proud than a good housewife. Her home may be spotless, her children neat, but she is too often lacking in those little qualities which betoken good management, a failing which belongs to the age more than any one class. Good housewives usually go in colonies, for the slattern is an affront to the tidy, and a clean and particular woman cannot long endure bad company, yet it is surprising how the sloven can improve when set an example by new neighbours. Certain canons of good conduct must be followed ; washing late in the week without good cause is a deadly sin, and clean curtains are obligatory at certain seasons. The Saturday clean up is as universal as its evening tubbing of the children, but nowhere can it be so thorough as in Cornish fishing villages. Every self-respecting woman has an extra clean after an absence or before highdays and holidays, extending perhaps to eradicating the grass growing between the pebbles of the path. Gossip is the besetting sin of the cottage woman ; at all times and in all places she cannot resist its insidious attraction. Too often ill-natured, many disagreements and not a few tragedies can be traced to tattle.

Happy as a rule is the lot of the country child, playing whenever possible in the open air and looked after by kindly neighbours when parents are ill or in trouble. Boys do not have such a hard time of it as when they went to work very young, although proud enough when promoted to some small job out of school hours, taking round the milk, delivering groceries, and light work which modern legislation will take away. The family do not now accompany their mothers into the harvest field to glean, but they often have tea with father under a hedge during haytime. The average labourer is fond of his children ; in summer they go to meet him, running happily by his side, or proudly accompany him to the allotments. In one village they are carefully brought up, obedient and tidy ; in another completely out of control, ragamuffins in conduct and speech and clothes.

Their games and playgrounds reflect the seasons, and the girls are still fond of the ancient singing games, which vary slightly in different districts and being handed down by word of mouth, often retain sound without meaning. Cricket and football of a sort and ball games generally are the boys' favourites, and the strip of green at a secluded cross-roads frequently serves as the cricket pitch of the nearest cottages. Games topical of the moment always have their attractions, and the younger children delight in playing scenes from the lives of their elders. The evening is the children's play time, their shrill cries and laughter revealing the village long before it can be seen. Sometimes the chant of childish voices leads to the discovery of a schoolhouse in a quiet nook, a centre for surrounding farms and hamlets, though often the children must walk long distances over the hill and through the wood to the nearest village.

A cottage not only tells the character of its inhabitants, but—— before the shortage of houses at least—reveals their social status as well, for each class of the community has its grades, definitely recognised by those who belong to it. A parlour, that holy of holies entered only on great occasions, confers a dignity on its possessor too seldom understood by superior persons who fail to recognise its influence. A hall, be it only a diminutive lobby screening off part of a living room, rejoices the heart of the cottage woman and makes up for many inconveniences. Both are little things which raise self-respect and make the cottager think he is not a being beneath notice. Windows may be small and so blocked with flowers the tiny half-casement cannot be opened, an affront to hygiene no doubt, but in summer the countryman spends most of his time out of doors and in winter he prefers a close atmosphere—it's cosy.

It is unnecessary to enlarge on the cottage for it already possesses a literature of its own. The garden is the joy of the average cottager, if merely a narrow strip of earth at the foot of the walls, gay with the homely blossoms beloved of his simple mind. He prides himself on the cultivation of some special plant, and when he desires to extend his vegetable patch unduly his good woman may say " I tell the master I must have my flowerses." Some cottages are grouped round a pebbled yard, but whether yard or narrow court, up which so many are cunningly hidden, some flowers are nearly always in evidence, especially in the

THREE GENERATIONS, DEVONSHIRE.

COTTAGE WINDOWS AT PETWORTH, SUSSEX.

AN AGED COUPLE
OF WIGMORE,
HERTFORDSHIRE.

" Blest who can unconcern'dly find,
Hours, days and years slide soft away
In health of body, peace of mind."

kindly climate of the south-west where the stone walls sprout blossom and greenery.

The pump stands near the cottage, at the top of the court maybe, or a windlass well serves several, but they are gradually giving place to the stand-pipe, utilitarian if unbeautiful, and the new houses possess their regular indoor supply. Where springs are numerous, lack of water cannot be put forward as an excuse for dirty homes, yet a few village yards are as slummy as any in the towns. In well watered-country, wayside wells beneath the hedge abound, with troughs and ancient conduits, and in Devon and Cornwall the pot-water or runnel beside the road, fortunately now seldom used for domestic purposes except in times of shortage.

Although the tendency is not so strong as it was cottagers still cling to their old homes and their native village. However, a change has its attractions, but except in the decaying townlets and villages opportunities for moving are few and when a cottage falls vacant a sort of " general post " takes place. The older folk usually prefer to remain where they are, generally to keep near an old crony, and the haunting fear of being turned out embitters old age. " I was born in the room upstairs and I hope I may die there " is an expression frequently heard.

Pathetic is the grief of some aged man and woman, too feeble to look after themselves, against whom an ejectment order has been made, and gallant is often their fight against fate. An old man of 73 told the magistrate that a doctor had only attended him for a scalded leg ; his cottage could not be insanitary as he had lived in it all his life, as had his father who lived to 96 and his mother who died at 89. They had reared ten children on eight shillings a week, and there had been no case of sickness in his family for sixty years. The sanitary inspector is the constant dread of such folk, especially when they own their dwelling. The squatter who has built a shanty on a bit of wasteland always goes in terror of having it destroyed ; in some parts where rough common abounds many still believe that their claim to ownership is incontestable if they can erect a dwelling and get a fire burning between sunrise and sunset.

To be removed to the workhouse, even in these days when that institution is more of a home of rest than the prison it was, is the last misfortune that breaks the cottager's spirit and brings about his death far sooner than the hardships of a solitary though independent existence. Our landowners and farmers are reluctant to turn out those who have served long and faithfully, and many are the pensioners living at a nominal rent and given some small sum for " services " which exist in name only but maintain self-respect.

The lot of independent workers who have outlived relatives and friends is far harder, but the real tragedy comes when they are neglected by those to whom they rightly look for help. Comparatively prosperous children for whom parents have worked hard not infrequently refuse to contribute towards their maintenance, callously allowing them to go to the workhouse. Such was the fate of the old man in the illustration. Years previously, having been deserted by his grown-up children, he had joined forces with the old lady, similarly neglected, and together they lived quite happily, although unmarried, until crippled by rheumatism, the chief foe of all agricultural labourers, over eighty and past work, his children refusing him the modest independence they could have provided, the poor old fellow spent his last days in the workhouse infirmary.

A few happy places boast that no one need want within their boundaries for generous inhabitants have left so many charities, albeit small ones, but there is sufficient for all in poor circumstances. But alas ! although modern government may have raised the general level of living, officials have no soul and little conscience. Charity Commissioners often descend on such villages, declaring a developing neighbour stands more in need of money, so ancient doles are ended and the funds pooled. Even almshouses, which provide so many aged folk with a comfortable home, occasionally suffer in this manner. Some of these almshouses are famous, both as buildings and for the romantic story of their origin ; many are ancient foundations with accommodation for a number of persons of both sexes in a building of picturesque appearance ; not a few consist of merely one or two commonplace cottages maintained by a comparatively small sum left by a person of quite modest means, or a row of modern erection, usually on the outskirts of a country town.

A romantic story attaches to a couple of simple cottages at the entrance to a sequestered village. On a stone above the doors is the inscription " All things come of Thee and of Thine own have we given Thee." Nearly three hundred years ago a servant of a neighbouring landowner ran off with a bag of money ; two others went in pursuit, all being mounted. The fugitive finding himself being overtaken threw away the money, and still leading, passes out of memory. Close by lived a farmer, and the story goes his sons found the money. At any rate he afterwards became prosperous and among his charities were the almshouses, the text on which is supposed to be a reference to the lucky find.

It is as easy to criticise the cottage folk of England as it is to praise their homely virtues, their quiet acceptance of the everyday tragedies of life, their resignation in adversity. Their curious beliefs and their quaint superstitions, generally, have a solid foundation on the customs and observances of the past and on that age-old folk-lore which provides the peasant with traditions not always recognised by his critics. Also they have a simple civilization of their own which has been handed down for countless generations, guiding their conduct and ordering their ways.

Times are changing and cutting them adrift from these old methods ; they are halting between the ways, they have neither shaken off the old nor taken on the new, so they are puzzled, dissatisfied, for having lost much of the good as well as the bad they are not yet alive to the advantages of the present. Formerly their simple wants were largely supplied by themselves out of their surroundings ; now they must purchase what they require, and often what seems improvidence is really lack of understanding of the right use of money. Unfortunately, prejudice, one of the great failings of the countryman of all classes, no doubt born of a self-contained existence, often prevents them making the best of what they have. Easily aroused and with difficulty eradicated, it usually takes a personal twist ; allied to self-sufficient independence it causes the rejection of counsel and instruction that would add to comfort and further prosperity.

Like the rest of us, cottagers have their vices, very ugly ones because vulgar and fundamental, which most have not the wit to hide. At the same time it would be foolish to accept an open countenance or furtive air at its face value. Their morals are often of a somewhat primitive nature, and evil frequently lurks in the most beautiful cottage, but some possess a very severe moral code of their own. What appears untruthfulness may be a clumsy attempt to please, or cautious acquiescence, and some of their transgressions are the result of old time customs. More than lenient towards the poacher and him who is not too strictly honest towards those higher in station, it is, perhaps, because they have a lurking remembrance of their rights in former days and the advantages they lost under the Enclosure Acts, combined with a hereditary memory of the harshness of the Forest laws. As a rule they are honest enough towards one another, and a lapse in this respect is often regarded as a crime as difficult to palliate as taking another man's wife. Rough and rude they may still be in their enjoyments, and their failings may be open for all to see, but hardly one is without some sterling virtue and the lives of many an idyll as fragrant as the nook in which they live.

A PEACEFUL TWILIGHT TO LIFE'S EVENTIDE.
THE TRAFFORD ALMSHOUSES AT PEMBRIDGE, HEREFORDSHIRE.

THE SUSSEX POTTER.

THE SHOESMITH'S YARD—LYME REGIS.

VI

THE VILLAGE CRAFTSMAN.

TRUE craftsmanship, the joy of the worker in the skill of hand and brain and in the finish of his handiwork, is nowadays more characteristic of the leisurely atmosphere of the country than the turmoil and haste of the town. The country worker has more variety in his tasks and must constantly exercise resource in coping with the unexpected and unfamiliar. The very seasons bring change, and as he is judged by results rather than pace he is encouraged to take pride in the execution of his task. He is the descendant of those unnamed artisans whose work is the monument to their skill, a rich inheritance to every Englishman who loves his country. Formerly the village craftsmen worked together for their native place or travelled in bands, as the church carvings in some districts suggest. What is better testimony to the old workmen than the village of Chipping Campden with its planning and buildings of different centuries ? Influence is so strong its modern inhabitants retain much of the taste and skill of their ancestors.

In the past, village and townlet were self-supporting, recruiting artisans from their own people, but the development of cities and the growth of factories compelled numbers who would gladly have remained to seek a living elsewhere. Although many industries have definitely migrated to the towns some of the oldest are so bound up with rural life they cannot be transplanted, and others provide a modest living for the few who prefer employment under conditions that please them to comparative wealth obtained in uncongenial surroundings. Everywhere we come across traces of the industries carried out in the country before steam power gathered workers into manufacturing centres. Obscure villages are known to fame because they gave their name to a particular kind of cloth, and a few little towns still flourish quietly on the industry carried on for centuries. Some have developed with the times, factories taking the place of the cottage workshop, and in a few the old employments are carried on more or less exotically, fostered by some guild or semi-philanthropic association.

When we speak of the village craftsman we think of the master workman rather than he engaged in some manufacturing process, although

the one melts imperceptibly into the other.   They are not so numerous
as they were, as the decay of agriculture brought stagnation to many
trades, and machine made articles, being cheaper, supplanted those
produced locally.   In the main the village craftsman is still an independent worker ; master and man are there, but the man is probably son
or friend.   The workshop is often a commonwealth of labour, each
doing his appointed share to the best of his ability, yet willing and able
to give a hand to the others when occasion calls.

First among the village workers comes the blacksmith, the tinkle
of whose hammer is one of the most typical of country sounds.   Few
can pass a smithy without stopping to glance inside, and schoolboys are
always anxious to take a turn at the bellows.   The smith has figured in
song and story from ancient times, his calling being one of the most
necessary and versatile of the countryside.   Unhappily, he is neither so
busy nor so ubiquitous as he was ; he has fewer opportunities of making
himself the clever artisan of the time when he fashioned the simple
agricultural implements.   Then, he and the farmer would lay their
heads together and add some improvements to those in use ; now,
machinery is so complicated makers usually supply spare parts, the
smith being required to undertake only simple repairs.   Formerly the
fame of those skilled in such work was noised abroad, and more than one
of our big agricultural machinery firms had their birth in the humble
smithy.

The smith, too, forged the weapons, the legendary Wayland Smith
being the ancestor of all who take especial pride in their work.   When
a peasantry aroused by wrongs or changes with which it disagreed sought
to assert its rights the smith gave assistance by beating peaceful implements into make-shift weapons, and not so long ago he made the
ironwork used in cottage and small house by the local builder, often as
artistic as durable.

Where small farms are numerous and the horse indispensable, his
forge is seldom idle ; he is equally ready to shoe a horse, tinker a cycle,
or make some special piece of wrought-metal for builder or housewife,
and just before haytime and harvest a medley of implements awaiting
his attention occupy odd corners near the smithy.   Maybe he is something of a wheelwright as well, and when the motor was introduced he
quickly gained a working knowledge of its parts, occasionally to such
purpose that his forge blossomed into a garage.   Now and again, perhaps,
among the cast shoes hanging on the wall is one bearing a name, which
tottering " grandfer," never so happy as when haunting the scenes of

KNIVES TO GRIND.

A LESSON IN BASKET MAKING.

A SLACK TIME FOR THE WHEELWRIGHT——SUFFOLK.

MAKING RUSH SEATS FOR CHAIRS.

his former labours, points out with pride as having belonged to a once famous racehorse shod by his father.

The coach-builder and wheelwright, too, may have relics of the past mouldering in the loft over his extensive workshop, a post chaise or a stoutly built carriage with ingenious contrivances under the seat or behind the cushion which added to the comfort of the traveller. He is another craftsman seldom called on to do work commensurate with his skill. Gone are the days when the lumbering family coach was brought to him for repairs or a fresh coat of paint, no longer is he called upon to overhaul the carriage, for the motor car has supplanted that, and the stables of the smaller gentry are mostly empty or occupied only by the cycles of the family.

Though many a wheelwright's shop is closed, here and there an old man with a reputation has more work than he can easily get through. In his yard and outside his shop stands a collection of vehicles of all sorts, from the light trap to the clumsy " muck cart " or lumbering waggon, some requiring a fresh coat of varnish and renewed cushions, others so badly used they must be almost rebuilt. All interested in country crafts can linger in the wheelwright's shop for days, watching the shabby tradesman's cart transformed into a smart vehicle, the gradual evolution of a waggon, the building up of a wheel and the shrinking on of a tyre, or the making of a wheel-barrow. Here will be learnt the properties and uses of different woods, and the special build of vehicles and embellishments favoured by the locality.

As the village trades are passed in review the same story of changing times and falling fortunes is brought to notice. Cheap ready made clothes killed the working tailor. Much of his work was done in the farmhouse ; taking his material and tools with him he sat on a table in kitchen or odd room repairing old clothes and making new, usually being paid by the day. The farmer's wife supplied the necessary patches, thread, odd buttons, &c., and provided meals. She was often kept busy attending his wants, and the story goes that when a much harassed housewife was told by her dairy apprentice that the tailor was asleep on the table she cried " Hush ! for heaven's sake don't wake him. I've had plague enough with him already ! " The more prosperous tailor worked in his cottage, visiting only the larger farms and houses ; he fitted out the game-keeper and hunt servants, and on red letter days the squire might order a suit. He still exists in the small towns, but the village knows him no more.

As the blacksmith is always the prototype of strength and up-rightness so the shoemaker stands for sturdy, not to say cantankerous, independence.   In a certain small town not much more than a village is a shoemaker of traditional type, always ready to scent a grievance or assert a principle.   Upon his shop front he daily displays annotations of the day's news written on sheets of foolscap, and many can point to a cobbler who asserts his views to those with whom he comes in contact, though he may not publish them for all to see.   Nearly always a strong politician he may even be a poacher from principle rather than inclination, and frequently a local preacher, often of some renown.   Though of bigoted views he is generally upright in character and a conscientious craftsman whose boots may be relied on.   As a rule he is a leisurely worker, the making of a new pair of boots taking as long as the erection of a house, for he lingers over them lovingly.   He, also, went round visiting outlying hamlet and farmhouse, executing repairs on the spot for which he received a daily wage.   The village cobbler finds less work for his skill, but boots must be mended and the requirements of the labourer in the way of nails and plates are best understood by those to the manner born, so he is usually busy on a bench outside his cottage in fine weather.

It is a pity he no longer relies on local leather, for the tanner has almost disappeared, though formerly a man of no small substance in country towns ; the disused tanyard sometimes remains, or its memory is kept green by the name of court or miniature square.   The saddler is still there, often a clever artificer in leather, though his shop is not confined to the sale of goods for horse and driver.   Perhaps we live in more humanitarian days, as the whip is seldom in evidence, especially those long in handle, bound with brass, whose lash reached the leaders of a team of six and cracked like the report of a pistol.

However, horse trappings are represented according to the fashion of the locality, from ear-caps of many colours to the most complete set of bells, face pieces, and amulets.   These are the property of the carter not the master and are of interest to the antiquarian, for their use goes far back in the ages ; some may have a pagan origin of which the crescent and the star are the most popular, while others are the special pieces of certain trades.   The huge flaps that stood up behind the collar, giving useful shade in hot weather, are becoming uncommon, and seldom does a carter invest in a net to cover the back.   Bells, once so proudly worn by the leader of the pack-horse train, are chiefly used on the gaily trapped horse turned out for the show, but the sun still flashes on crescent and disc as the Shire plods along the highway.

THATCHING.

"Thatcher, Thatcher, thatch a span,
Come off your ladder and hang your man."

HONITON LACE MAKER.

"The faster I work it'll shorten my score
But if I do play it'll stick to a stay,
So heigho! little fingers, and twank it away."

THE CARPENTER

" The village cobbler is still an institution, and he
has a considerable number of patrons."

Gone is the village clockmaker, another craftsman who has fallen before the double assault of cheapness and foreign importation. A clever mechanic and sometimes a genius, he has left behind sundry ingenious models and contrivances ; he made the round of the neighbourhood, winding and repairing clocks or doing odd jobs which required a delicate touch and mechanical knowledge. The tinker has not deserted his rounds, but trade is not what it was as pots and pans are too cheap and flimsy to be worth repairing, and grindling ill-used knives is not very profitable.

The miller is a master craftsman who too often finds the times out of joint—even war control brought more trouble than prosperity—and the millwright is difficult to find. The big steam mills at the ports deliberately set out to kill the little local mills, and modern taste prefers the flour ground by the roller mill. The well sinker would hardly flourish if he had no other work, for the civil engineer renders him unnecessary, but the water diviner often scores over science. Deny his special powers if you will, argue as you like, but if experienced in hunts for water you must have heard of the water diviner who has succeeded where the geologist failed.

A carpenter in his paper cap is a rarity, but there comes to memory a typical carpenter's shop beside his porch in a Cornish village, its large bench standing in the middle of the floor littered with shavings, tools lining its walls. Only for meals did he doff his square paper cap, and although it is uncommon to find such a figure and surroundings typical of old English pictures, the carpenter himself flourishes, often being builder as well, and certainly undertaker, three trades that go together in rural parts.

Allied with the carpenter's calling are a number of occupations, many carried on in the open air and purely local. Thus, in South Buckingham rows of chair legs leaning against the cottage railings or stacked in piles to dry are village characteristics, and in the Midlands is carried on hoop making for the barrels in which crockery is packed. Ash rods, sawn into lengths of various sizes, are split with a cleaving axe and shaved to the proper thickness with a draw-knife. Steamed to make them soft and pliable, they are bent into shape round a " horse " and wired into bundles for the barrel factories.

All who wander in the woodland by-ways can speak of little known industries carried on in cottage workshops. The making of wooden hay-rakes is one example, wooden pitchforks for use in training stables another, while the turning of handles for besoms or the shaping of the wooden part of brushes another. Birch and heather brooms are made where the raw material grows and a market for the finished product exists, and a few villages in Surrey and Sussex turn out wooden baskets or trugs, for which Hailsham is perhaps most famous. The white-cooper is a craftsman requiring some finding, for churns, wooden tubs, and so on are mostly made in factories, but very occasionally a maker of wooden bowls and the like earns sufficient for his needs. He is a skilled worker indeed, who uses many tools and is learned in the properties of woods. Most of these industries are notable for the results that are obtained by the clever handling of quite simple appliances.

Children may no longer sing the old rhyme to the thatcher; he has no doubt declined in popularity with the reduction in his numbers. However, his occupation has by no means gone, and within the past few years his labours have been much in demand. Methods of thatching, especially in the manner of finishing off and decorating, differ according to the locality, as to the names and shapes of tools to some extent. The old time thatcher loved to give his work an artistic finish, cocking the thatch into a peak or carrying it in swelling curves over dormers, putting a line of zig-zag hazel rods along the ridge, following his own taste and the fashion of the neighbourhood. For the best thatch un-bruised reed is required, the veteran worker demanding flail threshed straw if it can possibly be obtained. Usually it cannot, so he selects a good sound sheaf containing straws of even length and seizing it by the butt beats it across a barrel to shake out the grain. Then it is hung up and combed to get rid of the small and broken straws and carried in bundles or "yelms" to the roof. For thatching stacks less care is exercised but "any old straw" is of no use. Formerly, in some parts a little rye was grown to provide straw for thatching but wheat is now generally used, except in East Anglia where true reed is common and in parts where sedge grass takes its place. The methods of preparing straw for thatching vary and usually it is wetted to make it lie better.

There is a touch of magic in the way an old roof becomes transformed into a smooth slope of gold without sign of the spars that keep the thatch in place. Starting at the right, the thatcher places the bundles of straw up the roof beside the ladder and works from eaves to ridge. Shaking it out he lays it carefully at a slight slant and pegs down the spars laid across it; then taking his bat he beats it upwards in the direction in which it is laid, driving it firmly under the spars until unevenness disappears. The next layer covers the spar, much as one tile overlaps the one below, and so he proceeds until there is a smooth even surface to the ridge, which is finished off so no rain can penetrate and the eaves trimmed with a sharp hook.

The foundation or lower layer of thatch is kept in place by tarred twine threaded through a large needle and passed round the rafters with the help of an assistant below, and rough patches are held by ropes of twisted straw and hazel pegs as in the case of ricks. A good covering of thatch will last twenty years, and some old cottages possess a roof four or five feet thick. Roofing a building is paid for by the square yard, the rick worker by the stack, the farmer providing straw and pegs, carefully preserved when the stack is demolished.

As one travels through the villages of England some country craftsman will be seen at his task, one almost as old as time, another the product of modern progress, but so numerous and varied few know them all. Here and there the local potter earns sufficient for his wants, but he has largely lost his trade to Staffordshire, though where suitable clay abounds pottery making is essentially a village industry if the solitary potter be uncommon. From an art class started in a cottage in a well-known corner of Devon sprang indirectly the art pottery, useful and ornamental, purchased by thousands of holiday-makers, while at the severely practical end of the scale is a little kiln hidden on a Buckinghamshire hill-side famous for bricks for bakers' ovens.

The stone mason is a master craftsman confined chiefly to neighbourhoods where quarries exist ; if he possess local fame, he makes the head stones for graves within a wide radius, happy when relatives allow his pet fancies to have full play. Where rushes grow freely the maker of rush chair seats may ply his trade but usually he is only an itinerant repairer. The basket-maker is another who may be seen at work beside his cottage when withies provide him with the raw material, or the only sign of his calling may be a plate on the door and baskets of different kinds hanging on the railing before it. Where the angler congregates the fisherman and tackle-maker are encountered, identified by a creel, a mouldering fishing rod, a winch, or flies that hide coyly among the geraniums in the window.

Mention must also be made of the industries carried on entirely by women, but here again the number is less than when the cottager spun the yarn and wove the cloth. Here and there such industries have been revived, and many trades are carried on partly at home, partly at the factory, glove making for instance. In the seaboard villages the knitting of jerseys is an occupation that fills every spare moment, and does not hinder gossip, while in West Dorset the women sit at door or in garden busily " braiding " or net making. A wealth of romance is connected with the making of lace, but fashion is fickle, giving it many ups and downs.   Formerly it was made all over the Midlands and South-eastern counties, showing Huguenot influence, while further west the Flemings have left their traces. Some say its manufacture was introduced into England by Catherine of Aragon, and there are numerous explanations of the old term " bone lace."

Honiton lace sticks are generally quite plain, but Buckingham bobbins are often marvels of intricate workmanship in bone and ivory, many bearing the name of their original users.   Pillows differ, too, the Honiton type being rather small and flattish, easily carried on the lap, while in Bucks they are large and round, dressed in cloths, each of which bears its name, and supported on a wooden frame called the " horse."   In winter a glass globe filled with water was placed beside lamp or candle to concentrate light on the work ; the older workers also taught the young, four of these globes being placed at the corners of a small square table.   The workers sat in diagonal lines, the tallest nearest the light, so that the concentrated beam fell on the pillows of each.   Lace making still runs in families, the children picking it up quickly with little tuition, and at least one family of hereditary lace-makers in East Sussex uses for pattern a piece of lace given to a dead and gone member by the Huguenot who brought it to England long ago.

Button-making is only kept alive by associates of the Arts and Crafts Guild, though formerly a recognised industry in Somerset and Dorset, especially round Blandford where the girls had to make so many before going to school, " high tops " used on hunting waistcoats, " cart-wheels " and the rest.   Straw plaiting does not give employment to the large number of women it did a century or so ago, though a few workers remain where the best straw for the purpose is grown on the chalk.   Occasionally a cottager cherishes one of the splitting tools or the curious little wooden mills which pressed the straw flat.   Of old the workers bleached the plait which was sold by the score of yards, but this is now usually done in the factories.

When we see some of the beautiful things the old craftsmen have handed down to us, it is sad to think that the machine is supplanting the hand working in conjunction with the brain, and that the great factory is taking the place of the individual workshop.   The solitary worker, the master and his few assistants, are usually real craftsmen, whereas the factory-hand is the slave to a process, the unnoticed cog

CARRYING A YELM.

SELECTING THE STRAW.

THE THATCHER.

SPLITTING SPARS.

BORING THE UPRIGHTS FOR MORTISING THE HORIZONTALS.

THE FRAMEWORK USED IN SPLITTING.

NAILING ON THE BRACES.

PUTTING THE HURDLE TOGETHER.

GATE HURDLE MAKING.

in a big machine, doing the same thing again and again under identical conditions.   Some speak of the evils formerly attending many country industries, forgetful of the fact that they belonged to the age and were more rampant in the town.   Progress must be served, of course, but in doing away with the bad we have lost much of the good.   Scientific efficiency measured with a foot rule and timed with a stop-watch kills the spirit of individuality and enthusiasm.   The huge factory cannot foster true craftsmanship or maintain that close touch between master and man, which oils the wheels of toil and makes the daily round a congenial occupation.

VII

AMONG THE HILLS AND ON THE COMMONS.

SOLITARY as seems the way across the rolling range of hills or wide stretching common the experienced wayfarer knows how deceptive is this air of deserted aloofness. The sheepfold is hidden in a hollow of the downs, the troop of horses seem part of the heath on which they pasture, and the wavelike undulations swallow a man traversing a foot-path. Those who ramble amid their fastnesses at all seasons are acquainted with the many industries they foster and know that the very wastes minister to the needs of those dwelling beside them. Thus, happy is the village and hamlet on the fringe of a common which sometimes adds to the communal income, thanks to the charge for pasturing animals belonging to others who do not possess common rights.

Even a small strip of waste has its uses. The men find it handy for odd jobs of painting and carpentering as it is just across the way at furthest, with plenty of space and no fear of rousing the good wife's ire by making a litter in her domain. In summer, she brings out her wash-tub or at least spreads the clean clothes on the gorse bushes to dry and whiten in the sun, and somewhere, not always hidden, alas! is the ugly heap of household refuse. A few geese straggle about hissing at strangers, and the wealthy cottager's donkey mostly finds his own keep when work is over for the day.

Where upland commons extend for miles the cottagers and small farmers may earn their living almost entirely by looking after beasts sent from afar. Thus it is on Dartmoor where the moor-men take in cattle to pasture, a few isolated farms being ranches inhabited only in summer. When autumn returns, the animals are collected and driven to pre-arranged spots where they are handed over to their owners from the farms in the cultivated country.

Droves of half-wild horses are also to be seen on some of these ranges, notably Dartmoor and Exmoor, Wales, and the New Forest. Shaggy and unkempt, they take care of themselves and hoof away the snow to get at the grass, or if a patch of succulent herbage is half-buried

A WELSH SAW MILL AT BETTWS-Y-COED.

DARTMOOR PONIES.

AT THE LIME KILN.

THE DEEP WORKING QUARRY.

"Flowery heaths and open downs . . . .
Where gipsies camp and black-eyed girls are seen."

A ROADSIDE QUARRY.

A LONELY HILLSIDE HOMESTEAD—ARKINGARTH DALE, YORKS.

FURZE FOR FUEL—FERN FOR LITTER.

in the gorse they reach it in the same manner. At the " drifts " or roundings up on Dartmoor exciting are the scenes when the active ponies are collected in the pounds for examination and branding ; some are driven away for sale at the annual pony fairs, not an easy task as they run like the wind and make nothing of a stone wall.

Such is the procedure more or less on most of the extensive grazing grounds, and those who own the common rights are very jealous of their privileges which have been handed down for centuries.   In such districts the pound is still an institution, for all stray horses and cattle are impounded until claimed and the proper fine paid.

Sheep are characteristic of the wolds and lesser grassy hills ; they fleck the whale-backed downs, they gather round the dew ponds, those wonderful and simple reservoirs which puzzle the unlearned, or crop their way along the ridge, the shepherd with his dog beside him standing aloof, gazing into the distance wrapped in thought.   But the sheep are ever in his mind ; they stray in the wrong direction, a clipped word, an unmeaning sound to the uninitiated, and the dog is up and off, rounding up the stragglers and marshalling the flock, which done to his satisfaction he returns to his master's feet.   Crossing the downs of Berks and Wilts a sudden sound of baa-ing falls on the ear as the pedestrian almost stumbles on a fold in an unperceived hollow ;   on the wilder hills a wailing cry calls attention to a solitary ewe with her lamb, hardly distinguishable from the ground on which they stand.   As one climbs the wall of the intake there is a scurry as the sheep sheltering in the shade scramble to their feet to gaze at the intruder.

On the ridges of some downland ranges are the ancient droveways along which sheep have been driven almost from time immemorial, but such flocks are seldom met nowadays, perhaps just before one of the lingering sheep fairs of Sussex or Berks.   Lambing time and sheep-washing are with us still, however.   On the lesser downs and amidst the ploughlands the farms are seldom far away, and in a convenient sheltered spot the lambing fold is made, close to the field where the flock is feeding. Lambing is an anxious season for the flockmaster of the higher hills where snow may bury ewes and lambs, or days of cold rain cause perhaps greater anxiety.   The washing and shearing when the half-wild flocks are rounded up—a work of no small difficulty even allowing for the intelligence of the dogs which search every hollow and every corner of stone wall lest some be missed—have not quite lost all claim to be regarded as a rural festival, but like harvest there is more business and less merry-making than of old.

When the men of the hills are not looking after the animals there is much else for them to do. Each season has its work, and some of the labours continue the whole year round unless weather interferes. Winter has not yet made its last stand ere the dwellers of the " moorish hill," as Leland would say, light the heath fires, swaling the moor as it is termed in some parts, to burn off the encroaching bracken and furze and heather so that the fresh young herbage can grow again. Simple enough, no doubt, but requiring skill and knowledge or the fire may spread beyond its appointed confines, destroying altogether instead of renewing.

To the upland farmer and cotter the moorland yields opulently of its riches, although to the more luxurious dweller in the plain it seems a barren place of little worth. The moor-man's crops are scanty and he has little straw to spare for his pony, few cows and pigs, but when the autumn comes he goes forth to gather the bracken harvest, stacking it in his barn or yard. Fuel he needs, too, but where trees are comparatively few and stone walls take the place of hedges he must find a substitute for the faggot and log. The furze, which makes a flame of colour round his home in blooming time, brings warmth when cut and dried for use on the hearth

On the hills where deep layers of peat are found he and his ancestors for generations have obtained fuel, and if peat proper be wanting there is the turf, cut into slabs an inch or more in depth and spread out to dry in the sun. Here and there, peat cutting on a commercial scale has been started, but unless good markets are close to hand, or the bed is on a low lying heath near a railway such enterprises have usually ended in failure. More substantial works exist on some moorlands, buildings for extracting naphtha from the peat, but the result has mostly been the same, though the moss litter industry of Chat Moss is flourishing enough.

To know the industries carried on among the hills you must follow the tracks and footpaths which branch off from the road at intervals. A well trodden path winding amidst the heather is a short cut to the saw mill whose shriek proclaims its presence though it cannot be seen ; further on is the main approach used by the traction engine and trucks, a " corduroy " road, mayhap, following a more level if more circuitous route beside which are stacks of sawn timber, built according to length and size ready for loading. Anon a faint cart track disappears over the rising ground ; follow it as it waxes and wanes, now an unmistakable track, now two wheel ruts lost in the heather, now the stony bed of a streamlet. It may be the way to an isolated farm or a gravel pit, mostly a tangle of undergrowth. Its size shows that formerly it was extensively worked, and the rusty spade, the recent marks of horses feet, and the newly turned gravel in the corner prove that it is used at long intervals.

Elsewhere the diggings are a hive of industry ; the hill-side or plain is cut into terraces where men dig and sift, and carts come and go endlessly.   Maybe a tram line connects with the railway down the vale, and at stated times a little engine puffs up with a train of empty trucks. Often the hill-side above the highway is honeycombed with abandoned workings whence flints were obtained when the road was constructed. Now almost hidden by wild growth they stand forth as naked patches, when drought burns up the greenery, and one or two may still be used on occasion.

Quarries are by no means the least picturesque feature of the hills, be they the little semi-circular pits used by the local road-mender, or the huge workings where derricks groan all day, and the crackle of crushers and rattle of sieves cease only during the dinner hour.   At certain times a buzzer sounds, when the quarrymen seek shelter, or stand sentinel on roads and paths bidding passers-by to halt, until the all clear is given after the stillness following the reverberation of the final shot and its accompanying rattle of falling stones—a necessary precaution as splinters of stones and occasionally large boulders are hurled long distances.

Changes of fashion in the use of material and cheaper foreign importation have ruined many of our quarries, where the deep workings are pools of water into which rotting ladders descend, with perhaps rusty cranes standing near.   Maybe a system of old stone tramlines show the quarry saw its busiest period a century or so ago, as can be seen near Heytor on Dartmoor where is a groved granite track with many branches round the workings from which the stone for London Bridge was obtained, being thus transported to the canal far below.   Hidden among the trees on a Somerset hill is a quarry from which was obtained the bluish slatey stone found in countless dairies, used as the pavement of streets, and built up into water cisterns so truly cut and fitted that not a drop escapes, but the era of galvanised iron brought about its ruin.   The owner, loth to see old workmen cast adrift towards the end of their days, leased it at a peppercorn rent to one or two who worked it on co-operative lines ; in one corner of the extensive workings they hammered and drilled and chiselled, fitting together with the love of the craftsman those wonderful cisterns which last for ever and in the hottest weather keep the water cool and fresh.   They paid their way, and it is to be hoped such enterprise and independence survived the lean years.

Some quarries are lost amidst a heath, and some are great holes in the earth, criss-crossed by wire ropes and so vast the workers appear like mannikins.   All quarries are fascinating to watch, the men apparently hanging like flies to smooth walls or cheerfully knocking away the ledge on which they stand, but few call more imperiously to the idler than those at Llanberis, with their gigantic steps of slate, on which miniature trains puff in and out of tunnels, rising above the lake.   Strange is the silence just before blasting times, when all activity ceases, until the shots having boomed out and echoed among the mountains, the syren calls the workmen from their hiding places.

Quarrymen like miners are largely a race apart but seldom to the extent of those of Purbeck and Portland who cling to their ancient customs so far as modern conditions allow.   The hills of Purbeck are honeycombed with little quarries entered by short shafts up which the stone is hauled on trucks by a winch worked by hand or a donkey. Some are upon the cliff close to the sea, now seldom worked, and decaying wooden derricks on platforms above sheltered coves show how the stone blocks were lowered into barges, skilful navigation being required to bring the craft into position under favourable conditions and impossible when wind and tide were adverse.

Often beside a quarry in limestone hills a range of limekilns represents another decaying industry, for lime is not used for agriculture to its former extent.   Still, the local builder must be served and farmers require a certain quantity, but their boys no longer fight and jostle for priority of turn as they did years ago when they raced to the kiln and home again.

Of all the industries connected with stone flint-knapping is perhaps the oldest and most romantic though confined to the heath near Brandon in Suffolk, not far from the prehistoric pits known as Grim's Graves. Some tell us this industry has been carried on since Neolithic days, but others say it cannot boast such continuity and that its prosperous era came with the flint musket.   Its prosperity rises and wanes with circumstance, and although the flint lock gun for savage tribes is no longer regarded as reputable merchandise it is whispered that another outlet for the knapper's skill is provided by the spread of knowledge causing many to desire an arrow head or scraper.   Flint knapping is another industry which descends from father to son whose skill is hereditary. Simple it seems as the worker deftly chips the stone into flakes by hammers of various sizes, but the novice, be he ever so handy with a tool, quickly learns it is not so easy as it looks.

Clay digging belongs as much to the plains as the hills, and no one can acclaim the appearance of the brickfield.   The bed of a prehistoric lake may yield an almost inexhaustible supply of clay for pottery, and as the years go on the coppices of birch and alder and willow and hawthorn are ruthlessly torn up, in their place rising skeleton scaffoldings and a network of wire-ropes working the pumps which creak and groan day and night.   Old workings are lakelets of unknown depths in whose waters the moorhen scurries and calls and rare waterfowl sometimes build their nest.   The preparation of china clay is confined to the hills of Cornwall and West Devon, not exactly picturesque though the surroundings may be beautiful.   Beside little pits in wooded glens old buildings have been adapted to modern uses ; a stream works the machinery which puddles the milk-white clay, afterwards run off into settling pits and when dry packed in barrels.   White is the landscape at such workings, great hillocks greyish white that arrest the eye for miles, and the streams run like milk.

Mining is seldom regarded as a country industry, but a few isolated collieries may hardly blot a fair landscape.   Other mines are no more out of keeping with their surroundings than quarries, such as those of Derbyshire, of Wales which provide the little gold now found in England, and those few still working in Devon where the rubble heaps are quickly clothed in green by the kindly climate.   A mine abandoned years ago may be worked by a little band of adventurers ; they rebuild the ruined leat, repair the wheel and dig over the tailings, their modest machinery being in the open or under rough sheds.

Some of the west country mine workings form a series of buildings descending a steep hillside, from the engine-house and shaft at top downwards through the crushing mills, slime table sheds and roasting ovens, to the platform where the ore was loaded, and finally at bottom the huge water-wheel that worked the pumps.   In former days the silver and lead mines largely financed England's wars, and some of the workings date back to Roman days.   Bordering a byway are tenantless cottages with gaping roofs, a big chapel stands in the middle of a waste, and a well equipped modern store is the most prominent feature of a cross-roads in the back of beyond.   A mysterious settlement indeed, until close at hand almost hidden in the brushwood the crumbling walls of an engine house reveal the story of the place.

The iron mine seems to have little in common with the heather broom, but the old fashioned besom has a use in the steel works, and the cottagers of the Yorkshire moors would earn little if they depended solely on the sale of their brooms for domestic use. The industry runs in families, the regular makers producing a better article than the nomads who also turn their hand to it at the proper season. There is more in the making of a heather broom than meets the eye ; the heather must be of the right growth, cut with a special scythe or sickle, and left to dry before being stacked at the cottage workshop. With careful selection the heather is built up in the required shape and gripped in a vice while the " laps " or bindings of ash—the preparation of which is an art in itself—are adjusted and fastened, the handle, a pointed stake, being then thrust through and spiked in place.

At the edge of the wood on a slope which catches all the sun the pheasant rearer places his coops, a seasonal occupation of the game-keeper on a large estate, and adopted by others retired on small means as a congenial method of adding to income.

Another calling of the uncultivated borders of the wood and hilly wastes is that of warrener, of ancient lineage indeed but seldom repre-sented to-day. At one period the warren was an important adjunct to an estate, and on the few that remain the warrener finds plenty to do in the season, not the lightest of his tasks being to prevent unauthorised toll being taken of his conies. A warrener of another kind is not necessarily confined to the hills ; more commonly called the rabbit catcher he is usually one of those independent men of many jobs fitted to the season. He may appear from nowhere at the appointed time and depart into the unknown when his work is finished. Skilled in the trapping of rabbits and learned in the ways of ferrets and terriers he speedily clears farms where rabbits have become a pest, though it is whispered that like the rat catcher he is not quite so thorough in his methods as St. Patrick.

Besides the regular occupations carried on among the hills the various harvests mark the passing of the seasons. The cutting of bracken and furze by no means exhausts them. Some moorland cotters add a little to incomes by looking after the skeps of bees sent to them in order to ensure a supply of heather honey. Who that has wandered over moorlands in summer has not seen the groups of families and children armed with a motley assortment of receptacles gathering " worts " (whortleberries), and incidentally staining hands and faces livid purple until appetites are satisfied ? In some districts the school holidays are fixed to allow the children to be free at the proper time, but within the past few years the price has been so remunerative that gipsies and others, always represented, have migrated in hundreds to the grounds. Blackberries and sloes, hazel nuts and acorns are also children's harvests. Maybe, too, some down is famous for its mushrooms which go to the early riser, or a little rill deep in a valley grows luscious watercress welcomed by town and village alike.   The knowing ones can find a market for plovers' eggs and in favourable localities a few turn an honest penny seeking geological specimens.

Such occupations and seasonal industries are followed by the nomads, the gipsies and their kindred.   Fewer than formerly no doubt, but their camps, from the big encampment of many vans to the tiny hut, are familiar enough where heaths and wild commons abound.   Bye-laws and school attendance officers have made their lives harder and more furtive.   Few speak well of the gipsy and his like, and being independent folk they seldom open their hearts to strangers.   Some wander from one end of the country to the other, but many never leave a well defined area and in time you become acquainted with their rounds and can tell where to find them almost to a week.   They know every sheltered nook and hollow for miles around, and by observing these the wayfarer can often obtain comparative shelter when overtaken by an unexpected storm.

The country constable finds it difficult to catch them napping as they know exactly when to expect him and get wind of a surprise visit in a marvellous manner.   No doubt they poach and do not put themselves out to find the owner of a wandering dog or straying hen, light a fire beneath the notice board forbidding them to do so, and occasionally keep the police courts busy over a faction fight.   They are not so illiterate as they were, but it is still quite common for a gipsy woman to ask the shopkeeper of whom she has bought paper and envelope to write a letter for her.   Nuisance though they are at times, it is doubtful if they are quite so bad as painted, and experience has proved that many are more courteous to strangers than others with more advantages.

The sheltered corner where the wood ends just below the swelling shoulder of the ridge will seem very lonely when it knows their encampment no more, and after they have finally left the heath it will have lost much of its picturesque beauty on an autumn eve, when the rising mist gives weird shapes of enchantment to tent and uptilted cart before which the twinkling fires burn.

VIII.

IN THE WOODS.

THOUGH the great days of our woodlands departed with the era of wooden ship-building the lesser crafts of the forest glade and coppice still provide congenial employment for those who delight in the open air and the skill of their hands. Unhappily, it is the old story of decay, for the machine makes more cheaply and foreign imports and other influences discourage the regular use of home grown timber. Forestry is a leisurely art, for one sows what another reaps ; it requires an altruistic temperament to sink money for the doubtful benefit of posterity. The advent of the iron and steel ship led gradually to the neglect of our woods, which, as the years went by, ceased to be hives of industry and became in the main silent and little frequented except in the game shooting season.

The crafts of the woods are not so diverse as those of the hills, being roughly divided into three branches ; those which concern the maintenance of the wood itself, the planting, thinning and felling ; those dealing with the uses of timber ; and the few occupations centred on the preservation of game. In ancient days the countryman battled against the wood rather than for it, but at a comparatively early date the importance of conserving the trees was recognised. The records of the Forest and Manorial Courts are full of fines imposed for felling trees without authorisation or at the wrong time. Tenants of the manor usually possessed the right of taking undergrowth for fuel and sufficient timber to keep their houses in order, but only so much as would not cause permanent damage. On one of the royal estates a tenant who was presented for having felled an elm was ordered to " have a talk thereupon with the King's officer before next court ! " The clergy were sometimes sad offenders both in the matter of tree felling and poaching, for instance in 1267 the Abbot of Rufford was charged with taking 483 oaks from Sherwood Forest but pleaded a former charter in extenuation.

Of old our forests had great influence on our prosperity and safety, an influence restored by the war. One of our most important industries, the working of iron, had its birth in the woods. This began at an early

THE STRAINING TEAM.

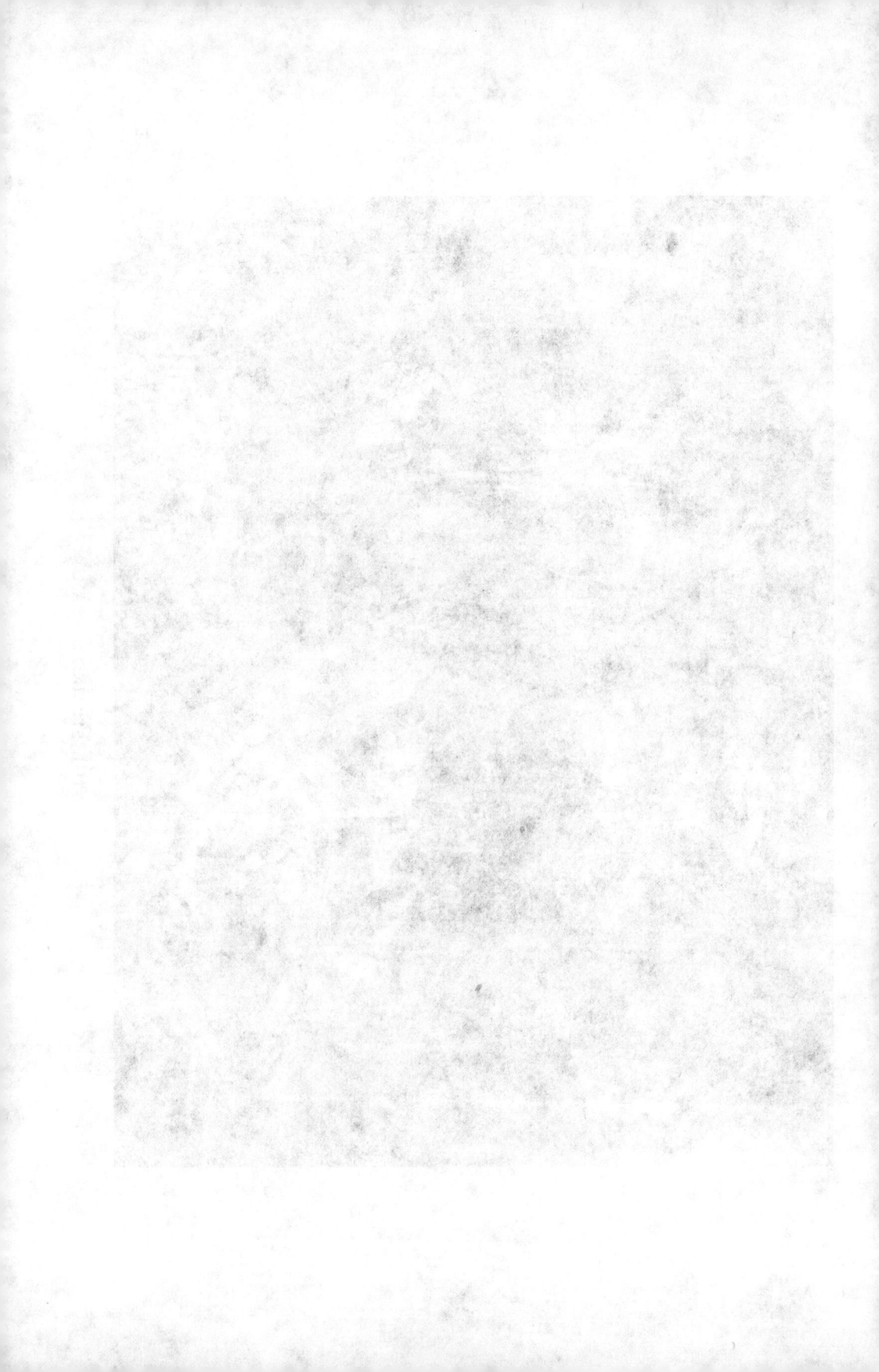

date as the remains of Roman workings have been found on the borders
of Sussex and Kent and in the Forest of Dean. The old ironmasters
became as wealthy as their modern representatives in the north, and the
great families of south-east England were not too proud to take part
in the industry, for it is largely a modern notion that an aristocracy
lowers its dignity by engaging in commerce. Relics of the woodland
factories remain in the " scowles " of Gloucestershire, and in Sussex
where the hammer ponds that stored the water for working the machinery
of the forges are still numerous, secluded pools almost hidden by the
trees that have grown since the furnaces were extinguished. Sussex,
indeed, can claim to have originated the modern armament industry,
for the first cannon cast in one piece came from Buxted; as the local
couplet has it,

> Master Huggett and his son John
> They did cast the first can-non.

Perhaps the law which allowed landowners to protect young growth
by the enclosing of land against cattle and deer for a specified period
after the cutting of coppice may be regarded as the beginning of forestry
as understood to-day. Later there arose " scares " concerning the supply
of oak for ship-building, when inquiries were held, stringent regulations
drawn up, and various methods adopted to encourage planting. Within
recent years the sporting rights have usually proved of more value than
the timber though it would be difficult to find a well-wooded estate
without its saw-mill. England possesses comparatively few woods of
large extent which are systematically managed for the production of
timber so that it is not easy to obtain an insight into all the arts of the
forester, from the selection of situation and soil suited to the tree to
be grown, the planting and thinning, to the felling of ripe timber and
replanting again.

Here and there timber is grown for a special purpose, but the
independent coppice worker, often a labourer who saved for years in
order to be his own master in a congenial occupation, has almost disappeared. He can still be found, more frequently in some parts than
others, and some are migrants who move from place to place in search
of supplies of the wood they use, but every year his number grows less,
and in the South of England, landowners who let coppice at the same
rate as cornland, keeping down the size of oaks which usually remained
their own property, have suffered in consequence. The tanner uses
chemicals instead of oak bark, and ready dressed leather is imported,
hop gardens are being grubbed or poles replaced by wire which is also

driving out the post and rail fence, the disuse of thatch means less call for hazel rods and pegs, and coal is the almost universal fuel. Thus, many a woodland workshop has fallen into ruin, and the pheasant struts where the faggot pile was raised.

However, the woodman is still familiar in many parts, and the idler out of doors can watch the felling of a great tree, from the clearing of the surrounding undergrowth and the ringing of the bark round the cut to the moment when the last blow on the wedge makes it stagger and then with increasing momentum fall with a crash exactly on the spot appointed. Busy is the scene as the branches are lopped off, the small ones trimmed and cut into poles and stakes, the spray tied in faggots, and the trunk itself sawn into convenient lengths. If an oak, and the bark is wanted by a local tanner, necessitating felling as the sap is rising, the ripping iron comes into play, and loudly cries the tree as it is stripped. Leaning against a rod supported on two forked sticks, the bark is left to dry while the woodman proceeds with other work. In many woods old and decayed trees are cut down and stacked in piles ready for the late season auction when country people buy their winter fuel. Often a tree is left for months where it fell, especially on steep ground where removal is difficult, but on a properly conducted estate it is soon removed by the waggon crew and their intelligent team.

The loading of the timber waggon is one of those great achievements which seem so difficult to accomplish, yet are carried out by quite simple means. Comparatively light larch poles and the like, unless very long, are usually transported on that type of waggon whose pole is fixed to the forward limber and slides through a hole in the rear one, thus allowing it to be adjusted for length. Stout rails are placed in an inclined position against the wheels, the chains adjusted, and the team, glancing round to watch progress, starts to haul the timber into position almost before the word of command. If the tree lies in an awkward position it is " chain rolled " into a better one ; the chain is wound round the trunk in several even coils, the end being attached to the horse, which on moving forward unwinds the chain and rolls the log.

When the bole of a forest veteran is removed a waggon is used, which allows the load to be slung beneath the axle. The pole is shorter and attached to a stout transom above the axle, which is in the form of a flattish bow ; when the pole is raised one side of the transom is nearer the ground and when pulled down the top of the transom describes a small arc. The limber is brought into position above the log, round which the chain is passed and one end fastened to the top of the transom. On the pole being pulled down the log is slightly raised from the ground and the pole is then fastened to the second limber; more chain is passed over the pole and under the load, allowing it to clear the ground so far as its girth permits. This picturesque timber waggon seems passing from the scene, its place being taken by the motor truck loaded by means of a derrick and carrying a bigger load more expeditiously.

Sometimes a patch of woodland ripe for felling is cleared completely for replanting, the woodmen being replaced by another set of workers whose business it is to grub the stumps, an easy task if they be small and the soil light, otherwise of more difficulty. The main roots are cut, and a wooden windlass erected in a suitable position. A stout chain from the windlass, which is turned by a horse, is attached to the stump, and as the chain tightens it emerges slowly, often with little jerks as the roots pull out and snap.

One woodland industry at least flourishes exceedingly although its prosperity varies with circumstances—the making of chair legs in the Buckingham beech woods where the plantations are systematically grown for the purpose. Each portion is felled in rotation when ripe for use, replanting being carried out when the " bodgers' "—as they are termed—camp moves on. Winter and summer the work continues in and around the huts; when working some distance from their homes the men live in the woods for most of the week, some being independent, others employees of a factory. The trees are sawn into proper lengths and split into billets, the bole of a felled tree forming the chopping block. Sitting astride the shaving horse with the billet fixed in front, the turner trims it roughly into the form of leg or rail by means of the draw-shave. When a sufficient number have been prepared he adjourns to the hut to turn them into finished shape by that primitive tool the pole lathe.

As simple as it is ancient, the pole lathe is admirably suited to the temporary workshop and is deservedly popular with the village and open air craftsman. The bench itself is rough and ready, as are also the spindles, and the working parts are simplicity itself. Beneath the bench is a large wooden treadle to which is attached a cord that makes one turn round the spindle and is fastened to a long flexible pole; the treadle points upwards when at rest, to take any weight off the pole. By depressing the treadle the article being turned is set spinning and the pole pulled down ; at the bottom of the stroke the foot is raised, and the pole, straightening itself, pulls the treadle up again and gives a spin the opposite way. The tool is brought into play at the downward stroke, and, practice making perfect, hand and foot work in complete unison, the tool being advanced and withdrawn quickly and deftly. As the legs and rails are of different lengths one of the headstocks can be adjusted, being held in place by wedges, and though the skilled turner largely judges by eye alone he puts marks on the rest supporting the tools as an extra guide. In the wood, the legs and rails, or stretchers as they are called, are built up in rectangular piles, in the village arranged in rows along foot of fence or wall.

The hurdle maker is perhaps the most universal of woodland crafts-men ; here he works all through the year unless hard frost stops his labours, there only for a season, according to demand and local custom. His hurdle, and perhaps the names of its parts, varies with the district. The wattle hurdle is easily and quickly made ; the necessary tools are few and simple—the curved frame with holes at equal distances in which the uprights are placed, and possibly only a billhook. It is good to watch a hurdler at work, sorting the wood and cutting the uprights into length until he has a number ready, leaning against anything handy, often a finished hurdle which he uses as a gauge, measuring with his eye. He sharpens the ends of the uprights, sticks them in the holes of the frame, splits the willow rods and swiftly laces them in and out, with a twist at the end to prevent splitting as he weaves in the reverse direction, finishing off by trimming any projecting ends with hook or light axe. Then taking the hurdle from the frame he presses it straight with his knee, on which he wears a leather pad.

The making of a gate hurdle is not accomplished so quickly, and the outfit of tools is larger but quite simple except that used for enlarging the holes into which the cross-bars are mortised. Well-seasoned timber of somewhat larger size than that used by the wattle hurdler is required, and a framework to hold the wood while being split. This may be described as a tripod with a couple of cross-bars nailed to two of the legs, one in front, the other behind ; they slope towards one another so that timber of varying thickness can be held loosely between them while being split. Sawn into proper lengths and the bark trimmed off, one end is pointed and the holes for the rails bored and shaped. The uprights ready, the rails are selected, cut to length, and flattened at the ends to fit the mortise. Then quickly and firmly they are fitted into the uprights and driven home, and the centre piece and two cross-bars

CARTING FIREWOOD IN BUCKINGHAMSHIRE.

"BODGERING" IN THE CHILTERNS—SHOWING THE POLE LATHE
IN THE HUT AND OTHER SIMPLE APPLIANCES.

THE CHARCOAL BURNER'S MID-DAY MEAL.

THE PILED HEARTH BEFORE IT IS COVERED WITH TURF—CHARCOAL
BURNING IN THE NEW FOREST.

BRUSHER MILLS, THE NEW FOREST SNAKE CATCHER.

THE WOODCUTTER—NORFOLK.

FIXING THE CHAIN IN POSITION.

THE WINDLASS.

THE UPROOTED STUMP.

IS DINNER READY?

GRUBBING TREE STUMPS.

or braces nailed on. The custom of the district or the idiosyncrasies of the hurdler show minor differences ; the cross-bars may project beyond the top or bottom rail or be cut off flush, and the rails may be pegged or merely fixed firmly in the uprights. The usual gate-hurdle consists of two heads or uprights, five rails, a centre-piece, and two cross-bars or braces, but larger ones are sometimes made for special purposes.

Post and rail fence making is not such a regular occupation on the woodland estate as it used to be, for wire is steadily ousting wood. The procedure is somewhat similar and though the timber is usually heavier the same contrivance for holding it is used, and the worker often turns from one occupation to the other. If the hedge carpenter works for himself he may use the smaller wood and chips for thatching spars, clothes pegs, and so on, but most of the waste goes for firewood.

The clog sole maker is probably best known in the north, but within recent years he has been seen at work in the south and west. Alder wood is most frequently used but birch sometimes takes its place. Sawn into butts or billets of varying thickness according to length so there may be economy of material, it is split and the clog sole fashioned by a single tool, a curious guillotine with a long handle. The other end has a hook engaging a ring fastened to the bench or block, and a toothed plate holds the unshaped clog in any desired position. The clogger picks up an unshaped billet, glances over it, holds it lightly in place, and quickly raising and lowering the handle shaves it into form. The finished soles are then stacked into dome shaped piles through which the air can circulate freely to ensure quick drying.

Hoop-making is largely a local industry as once again iron has supplanted wood, but in the Midlands and elsewhere a hoop-makers' camp may sometimes be found. Like the hurdler and others he wears a leather apron or pads of the same material in places where the wood rubs. Hazel or ash is his material, and axe, adze, hook, spokeshave, and drill or small brace and bit his tools, with such necessary adjuncts as the chopping block, the post for holding the rod while being shaped, and the horse for making the hoop itself. The horse is a flat round trestle of about the circumference of the hoop, which is passed underneath and the ends brought together on top for drilling a small hole to receive a wooden peg. Measured against a gauge for length, passed round the horse, the peg inserted, and the job is done. Sometimes the hoops are merely cut into lengths, split, shaved, and left as they are, suitable for casks of varying sizes.

You can see a hedge carpenter begin and complete a specimen of his art in the course of a morning stroll, but it requires a sojourn of some three days in the burners' camp to learn the mystery of making charcoal. Your charcoal burner would require some finding for he is not a common inhabitant of the wood nowadays, but he has not quite deserted the New Forest, and occasionally his hearth may be seen in Surrey and Sussex, in the Midlands, and elsewhere. The building of the pile or hearth is not so simple as it appears ; the billets cut to suitable lengths must stand upright with a slight slant towards the stout central stake, round which they are stacked loosely to allow circulation of air but exclude a draught.

When the pile is finished it is covered with grass over which sand or earth is shovelled and pressed down carefully to keep out the air. Next the central stake is withdrawn and the hearth lighted by means of brushwood inserted near the top. The hole is then covered with a sod of turf and the charcoal burner sits down for his vigil ; in windy weather he cannot afford to doze as should the hearth break into flame it would be spoilt, so turf is handy to cover such an outbreak, and if the weather be stormy screens may be erected. As the hearth burns it subsides slowly without breaking the seal, and the volume and colour of the smoke tell the burner when the charcoal is almost complete. Then water is thrown over the pile to drive the steam inwards and extinguish the fire, the covering is withdrawn, and the charcoal left to cool. Near the hearth is the hut of the burners, often a wigwam-like structure covered with turf and branches, for each takes it in turn to watch the pile.

All these use the wood as their workshop or temporary home, but there are a few who know no other dwelling-place, a nomad who lives the retired life of a hermit in a little frequented thicket, or a regular worker who prefers a semi-wild life. Such was the woodman, cunning in all the woodland arts and able to drive a straight furrow when required, who passed more than half his life in the woods of the Duke of Norfolk's Sussex estate. Then there was Brusher Mills, the famous snake catcher of the New Forest, who lived for thirty years in the most primitive of huts until the Court of Verderers ejected him. Sanitation and the constant company of his fellow men did not agree with him after his solitary existence, and, as often happens, he pined and died. He counted his captures by the thousands and had much to say on the vexed question of adders swallowing their young, but characters such as he do not flourish in these modern times in spite of their easy content, for those fond of the life of the primitive backwoodsman are generally rounded up by the authorities.

The monarch of the wood is the gamekeeper ; the wood is his kingdom, his palace the cottage in a glade or beside a road on its outskirts. Within its boundaries his word is law and he does not always welcome the woodman who may disturb his game. Regular hours he has none ; he comes and goes by day and night, to look after his newly hatched pheasants, to watch for a gang of poachers. In these days, when estates change hands so frequently, he moves about more and is not so often the old retainer he used to be, passing all his service on the same property until pensioned off in honourable retirement. He is a better informed man than he was and he is not quite so prejudiced. His gibbet does not show so many useful birds, but his knowledge of natural history is deeper and more accurate. He is learned in dog lore, and what he does not know about rabbits and ferrets is hardly worth knowing. His acquaintance with rural work is wide, as he superintends many labourers on the estate ; he can seldom call his day an easy one, but he feels quite content when at the close of the great autumn shoot he is congratulated on the fine show of birds, the manner in which he placed the guns, and the clever beating of the covert. It has been said that the qualifications of the ideal keeper are, " Knowledge, skill, perseverance, discrimination, firmness, order, courtesy, and enthusiasm," a lengthy list which can be accounted to his credit more often than might be supposed.

It is a curious fact that many of the workers of the wood are almost quite ignorant of natural history, and cannot tell the names of some of the commonest birds and wild-flowers, though their opportunities of observation are to be envied as their presence becomes so familiar the wild creatures pay little heed to them. A few, certainly, are experts if shy in imparting their knowledge, and numbers become wonderful weather prophets, foretelling rain with surprising accuracy and knowing when a fine spell may be expected after a period of inclemency. The wood may not appear to lend itself to such prognostications and few can say how they arrive at their conclusions, no doubt by instinct, automatically observing sounds and signs which have unconsciously become significant to them.

Perhaps the woodman possesses special qualifications for picking up such knowledge, as he loves the open air if anything more than the average countryman. Usually he has taken up his occupation because it appeals to him, and he rejoices in the skill of his hands, being quite content if it provides him with sufficient for his wants, and sometimes he has saved enough to become his own master in a modest way. So, satisfied with his lot and happy in his craft, he envies no man, for he has discovered the secret of a contented existence.

IX

CALLINGS OF THE COAST AND WATERSIDE

No true Englishman can resist the attractions of the sea for he would be denying his heritage ; to this day those who dwell within hearing of its murmuring voice often lament because their 'prentice boy has stolen away to answer its call. In peace and in war the Englishman struggles with the sea ; although he may not yet have mastered her he has harnessed her to do his bidding, and well has he earned the right to labour on the more placid river and find an easier living practising the arts which may be followed on its banks.

The great ports are full of hidden romances but they belong to the city rather than to the country. They are the gateways of the ocean as the little ones are the entrances to the narrow seas ; the former serve the nation, the latter a few towns and villages, at most a district. To their miniature wharves come the coasting steamer and sailing ship which travel to and fro with cargoes of purely local value. Busy as may be these little ports in a quiet way their life runs easily ; the men hoisting baskets of coal or transferring bags from farmer's cart to hold cease their labours to discuss the identity of a passing stranger, and the captain spends his leisure in the inn beside the quay. At low tide the vessels are careened to caulk a seam or receive a coat of paint, while outside the sail-maker's shed a new suit of canvas is being fitted or worn ropes replaced. Perchance there is a small dry dock in which work can proceed at all stages of the tide, or a builder's yard where the old-fashioned tap tap of the hammer has not been replaced by the ear shattering roar of the pneumatic riveter.

Our coasting ships face perils as great as those encountered by the vessels which cross the oceans ; they ply along ever thronged sea-ways, they have a dangerous coast under their lee, and the weather changes almost from hour to hour. Their voyages may be shorter but they spend nearly as much time at sea, for no sooner are they unloaded than off they go again. In the lesser ports within easy reach of the coal exporting districts the squat colliers with smoke-stack sitting on

THE EEL-CATCHER—BRANDON, SUFFOLK.

A COCKLE WOMAN OF THE EXE.

BOATBUILDING AND SHIPBREAKING ON AN ESSEX RIVER.

CRABBERS.

the rudder seem to make a continuous voyage to and fro; riding high in the water they have barely rounded the pier ere they return heavily laden to moor at their accustomed berth as the crane moves up to start unloading.

Sailing ships are seldom in the same hurry, and barges are more leisurely still; they like to have a look round to see if anything has changed while the crew depart for the "Anchor" and the captain goes in search of the consignee. Before villages up shallow estuaries and creeks, schooners and ketches appear at long intervals, settling down in the mud like squatting ducks as the tide falls and waiting to unship their cargo into carts which come splashing through the ooze; barges remain so long at grass grown wharves that their departure comes as a surprise.

Some of these ports deal chiefly in one commodity; their seaweed-hung weather-beaten staithes are white with china clay, or backed by ranges of red-tiled maltings. In many the sailor mingles with the fisherman, the timber ship from Norway lies beside the drifter amidst a huddle of boats that seldom go far from shore. At certain seasons of the year when the mackerel and herring fisheries are in full swing the fishing craft are supreme and the shout of the auctioneer is heard all day until the shoals pass on and sleepy industry reigns once more. The Scottish lassies who come to clean the fish for curing are synonymous with the herring fishery season in East Coast ports, and in the smaller ones of the south-west are the quayside factories in which pilchards are packed for export.

Within recent years a great change has come over the fisherman, for the steam trawler and drifter have almost ousted the smack. The skipper owner is fast becoming the servant of a syndicate, the crew sea labourers, and the local diversities in profit-sharing are dying out. All who love the picturesque processions of the sea miss the brown-sailed yawls and luggers that filled East Coast harbours, but Brixham—which claims to be the original home of the trawler—and other ports have not yet been quite conquered by steam, while the motor auxiliary engine is giving the smaller owners a chance.

Only by frequenting the coast at all seasons is it possible to understand how varied is the fishing industry, for methods local and peculiar will be missed unless observed at the right place at the right time. All connect certain places with special catches, such as Aldeburgh and Brightlingsea with sprats, but the sand eels of Teignmouth probably have more local than general fame. Lancashire and the East Coast are the homes of the shrimping trawler, whose net drawn to the mast-head to dry relieves the flatness of Essex estuaries, while mention of Cornwall calls to mind the huer on the cliff summoning the seine boats as the pilchards have come. The seine in all its glory may best be seen in the south-west, but nearly all round our coasts it is cast by boats just beyond the breakers and hauled in from shore, perhaps where rocks and boulders seem to make its use impossible. On some tidal rivers the salmon netters, those aristocrats of their calling, ply their trade. Shooting their nets nearly across the channel, they row ashore and wading in thigh deep draw in their catch, which is too frequently a sorry reward for labour.

Along little frequented beaches and before fishing villages the drift-net luggers and boats of the line fishers are drawn up, the catch often being purchased on the beach by hawkers who travel the country side. Stake nets are a feature of the gently shelving shores of parts of Wales and Lancashire and are still used for salmon where England meets Scotland on the north-east, and in the neighbourhood of Rye the ranges of keddle-nets may be seen. Oysters are more widely cultivated than some imagine, and 'tis whispered that Whitstable and Brightlingsea are not the homes of all that bear their name. That typically little port of the salt marshes, Emsworth, has a fishery, and in autumn the bosom of the beautiful Fal in Cornwall is dotted with the oyster boats whose crew of one or two twirl the handles of the windlass winding up the trawl.

The crabbers and lobster-men are widely spread and usually full of lament concerning bye-laws and havoc caused by careless inshore trawling. More plebeian and none too remunerative is the work of the winkle gatherers, and energy is displayed by the cocklers of Morecambe Bay whose carts are uptilted on the gleaming sands until the rising tide drives them home. Quaint are the women cockle gatherers of Stiffkey in Norfolk, with skirts tucked up, broad straw hats on head, and deep baskets slung across shoulder. On this marshy coast, where the wind blows direct from the Arctic, cockle gathering is for the strong in limb and hardy in constitution, although under sunny skies it is not unpleasant. However, even the mussel rakers of the Exe find it necessary to tie on their hats with ample kerchiefs, and at times a biting wind blows across the flats. The cockle may be a humble dainty, but it possesses a lore of its own, each district having its approved method of gathering and fashion of dress for the task, sometimes its own superstitions, its belief in the supreme perfection of its own produce, and its special manner of cooking and serving. The gathering of these varieties of shellfish is no small industry, and at King's Lynn are allotments devoted to their culture.

Conditions may be changing but fishermen's talk is still mainly of the weather and the tides which have more to do with their calling than seasons or clock time ; they still spend hours gazing out to sea, with intervals mending nets or perhaps tanning them in the boiler on the jetty. Some combine fishing with farming, while practically all fill in their time with the many odd job callings of the coast, and some serve in yachts in summer. Their beliefs and superstitions are dying slowly as they mix more with the outer world, but even now the fishers of the Chisel Beach follow the centuries old custom of making offerings to the sea at the appointed time in order to ensure a successful fishing, and rabbits, especially white ones, are not approved by fishermen in some parts.

" Who. whipped the hake ? " cries the man of Zennor who would arouse the ire of one from St. Ives. In that dim era when legends were actualities hake sadly plagued the St. Ives mackerel fishers, so hoping to improve their conduct they whipped with little rods the largest they could catch and cast it back into the sea thinking it would spread the news. Nestling under a jutting chalk headland is a village which perhaps finds fishing and catering for visitors less remunerative than its former trade of smuggling. Its neighbours tell a scandalous tale that when the hour of putting forth arrives its fishermen light a candle halfway up the cliff ; should the flame flicker 'tis too windy to go to sea, should it burn steadily then surely there is insufficient breeze to fill the sails.

One coastwise occupation calls up another as they are reviewed in memory, and a stroll along almost any mile of shore will show how easy it is to forget and how small is knowledge. Almost at once leaps to mind the lifeboatman, whose name alone is more eloquent than a description of all it means. Yet, really to understand, one must have watched the lifeboat disappear into the night in response to the dull boom of gun of distress or flare that barely pierces the murk, stood among the knot of watchers as the hours go by without news or sign other than the thin trail of rocket creeping up the curtain of the sky ; the grey dawn when all eyes are straining seawards, the eager questionings of new-comers who may bring tidings of landing along the coast, the rumour of disaster that springs from heaven knows where, and the hysterical cheer when one more keen sighted than the rest picks up the lifeboat in tow of the tug.

The work of the men aboard these fussy little steamers seldom receives its due ; associated mainly with towing sailing ships over sunlit waters or shepherding liners in and out of dock, few except the dwellers in the ports and along the coast know of winter work when they venture far out into the stormy ocean to bring disabled vessels to safety. Often the lifeboat itself is powerless without their aid ; only in tow can it leave the narrow harbour entrance into which the gale blows with full force. It may be difficult to return, and tug with lifeboat behind must steam up and down seeking a safe landing place, but the advent of the motor seems likely to ease the labours of both lifeboat and tug. Among the life-savers are the Coastguard, equally ready to man the rocket apparatus or climb down the cliff to rescue a tourist whose foolhardiness has brought him into danger, answer foolish or intelligent questions, or enlarge on the discomforts of their exposed dwelling.

The salvage men who seek to refloat the heavily stranded vessel or strip a wreck of fittings and valuables have no easy time ; their task may be quickly finished or last for months if they can only work in good weather at certain stages of the tide. The fishermen and surfboatmen do the same for the smaller wooden ship, but only ancients on shores bordering once famous roadsteads can tell of the heyday of anchor sweeping. Lower in the scale are the longshoremen who search above low water mark after storm and do not always report their finds to the receiver of wrecks. Little romance of the sea attaches to those who seek lost articles on holiday beaches, an occupation that occasionally brings good profits.

Egg gathering on the cliffs is unlikely to attract many except those to the manner born, such as the men of Bempton and adjacent villages in Yorkshire, who go over the edge and, without turning a hair, swing perilously at the end of a rope two hundred or more feet above the surf. Sand or shingle carriers loading the panniers of donkeys or ponies make a picturesque group in many Cornish bays, and perfectly in keeping with the scene are the seaweed gatherers of autumn ; their carts cast light shadows on the wet sand and the rude blast sends strands of weed flying across the beach and threatens to strip the patient ponies of their shaggy coats.

Lighthouse keeping is perhaps the loneliest calling of the coast, though that of the men on the powder hulks in sheltered bays and unfrequented creeks may be more monotonous. The lighthouse keeper has a companion or two, but if his post be on an isolated reef or sandbank his relief may be delayed for days if not a week or two, and those in a tower at the extremity of a far projecting headland see few strange faces.

Life goes more easily for those who earn their living on inland waters, but the coaster and the barge which have braved a hundred

ATTENDANT ON THE HERRING FISHERY—SCOTCH GIRLS WHO
COME SOUTH TO CLEAN FISH.

DREDGING THE THAMES AT ABINGDON.

A BROADLAND BOAT BUILDER'S YARD.

A CORNER OF THE FISH MARKET—NEWLYN, CORNWALL.

ON THE CANAL AT EASTINGTON.

LAYING THE KEEL—ST. IVES.

tempests in the open sea may meet their end in some landlocked shallow, for sudden squalls arise and the surrounding hills cause strange winds that baffle the mariner unaccustomed to local conditions. Wind and tide together influence the currents so that even the ferryman who knows the channel's every mood finds himself out of his course, especially when blinding rain and snow blot out the landmarks, instinct allied to experience enabling him to reach his destination. None know this better than the navigators of the little steamers carrying passengers and merchandise on estuaries which penetrate deeply into the land, stretching forth as many creeks as a hand has fingers. These form the chief highways, and the skipper crossing from quay to staithe or diminutive wharf at the end of a lane must be as alert as a bus conductor in picking up passengers lest he miss one waiting at a little used stopping place.

Shallow water sailors also form the crews of the picturesque Thames and East Coast sailing barges, which, loaded to the gunwale, put to sea with every confidence. In skill and contempt of danger the deep sea mariner does not surpass these men who seem to spend so much time in red roofed inns in Essex, Suffolk, and Kent, and may be met in ports all round the coast. The broad beamed Norfolk wherry, whose dark sails appear to be moving smoothly over the land when seen from afar, also makes short sea voyages. We are told its days are numbered and that it will soon be rare to see one propelled along a narrow cut by the long pole or dipping its mast on approaching a bridge. The horse-drawn barge with its brilliant paint is not without attraction, especially if the tiller be held by the bargewoman in her flowing sun bonnet. Often the crew take things easily, the bargee lazily lolling against the tiller, the horse which knows the work perfectly plodding on until the lock is reached, when it moves forward or halts as circumstances dictate without needing promptings from its master.

Monotonous is the life of the lock-keeper on the canal though there are places where floods may keep him about at all hours. On pleasure streams he is busy in summer, passing an endless procession of craft through his lock and superintending their passage over the rollers, shouting instruction in elementary navigation to novices who threaten to shipwreck their passengers, or making a river bounder behave himself. The task of the ferryman, his virtues and his failings, have been sung so often that it is no injustice to dismiss him without further remark.

Ship-building is regarded as an industry of tidal waters, but high up the creeks, on backwaters of the Broads almost hidden by the swaying reeds, on the banks of rivers, beside the placid canal, are little yards where smaller craft of all kinds are launched. On a slipway embowered in trees the graceful lines of a yacht grow slowly, boards of the rarer woods to embellish its cabin being stacked in the adjacent shed. Where the tide is hardly felt are yards accustomed to the building of smacks and sailing barges, with newly launched vessels moored next old ones awaiting refit. One yard may possess some small fame for its steam launches, and another destroys as well as constructs, firmly embedded in the muddy foreshore being hulks in various stages of demolition.

In one place the ordinary barge is constructed on a travelling carriage or cradle running on rails. This is lowered down the slipway into the water to float off its burden when complete, another requiring repairs perhaps taking its place. Nor must the builders of light river craft be forgotten, some renowned among oarsmen for their racing skiffs. Occasionally these small yards combine boat building with saw milling, especially in wooded country where the barges bring the trees and take away boards and posts, and not infrequently the mill works on after building has been given up.

Numerous are the tasks of those employed by the water side—repairing docks and locks, keeping in order weirs and salmon ladders, building up and refacing river banks at bends or removing the sand and shingle which accumulate in such places in fast flowing streams, for if this be not done the current quickly eats away the shore and may find a new channel. Mud must be dredged from the more sluggish river, weeds kept down, the hatch of the mill leat and the pit in which the wheel works regularly looked to, a length of the canal emptied now and again to facilitate repairs, and in low lying country those who watch the banks must never relax attention lest a weak place let loose the flood. In such parts too, the pumps must be kept in order, while much labour is employed clearing dykes and ditches.

In the fens and broads the marshman has his home. His occupations are varied indeed, for he turns his hand to any work the seasons bring. When winter comes he seizes his short bladed long-handled scythe and mows down the reeds, which are tied in bundles and stacked to dry. For recreation he goes out in his gun-punt for a night's shooting, though as a rule the wild-fowler, a disappearing class, is a race apart. He may catch eels or collect eggs, and in summer probably has charge of a herd of bullocks put out in the marshy fields, not such a sinecure as it sounds, for cattle will stray and may be stogged so must not be left entirely to themselves

An industry employing considerable numbers is withy growing and the making of baskets although all its processes may not be carried out in the same place. Where the willow grows freely its young shoots are cut regularly, once a year or at longer intervals, and either prepared on the spot or sent elsewhere to be dealt with. In almost any district where a river spreads into a marsh and willows flourish the withy cutter is busy in winter. On that great tract of fenland roughly known as Sedgemoor ground suitable for withy growing is leased for the purpose, and the young willows are tended carefully and kept clear of weeds. Once planted they yield a better crop each succeeding year, until when strong and hardy the grower need not worry about the weeds for cattle can be turned out to eat them without fear of doing harm.

The withies are cut, tied in bundles, and placed upright in pits of water for several weeks. By early spring the rind is soft and will strip easily. The strippers work in little groups of happy family parties or larger ones of women sprinkled with men not employed in the fields, but elsewhere men may predominate. Before each worker is a wooden post about thigh high, on top of which are two pegs of iron between which the rods are stripped, the withy brake by name. The stripper places an end between the pegs and pulls, bringing forth a clean supple rod which is placed against wall or hedge to dry. If buff coloured rods are wanted they are boiled before stripping, the dye in the rind giving the colour. Some cottagers make their own withies into baskets but most are sent away, and in unexpected places are wayside basket factories where withies are stripped and plaited.

Always something new bids the loiterer linger in the water meadows or beside the crooning sea, for a chance acquaintance seems ever waiting to initiate him into the mysteries of a new occupation. Now he learns the secrets of the fish hatchery, now he may master the details of successful watercress culture, and one day seeking unknown ways among the lesser uplands he comes on the waterman visiting the spring that supplies the townlet seen among the trees below. Amid the higher hills and mountains are chains of artificial lakes and great reservoirs with mighty dams that testify to the skill of the engineer but whose waters do not invite a sampling taste as do those of the sparkling spring.

Cast a fly in prohibited waters and the water bailiff pops out from the bushes ; visit the hamlet by the trout river and the local fisherman, expert in the art of tying flies and acquainted with every mood of his stream, offers his services. Roam about the high ground overlooking an estuary and watch the men who attend to buoys and channel-marks, and from the beaches bordering the narrow seas observe the pilot boarding the steamer, or from the pier watch the sunset enclosing a little harbour.   When inclination refuses to heed the warnings of duty that the hour for returning is overpast avenging fate may bring the rubbish tip with its evil smelling barges, or the odours of the fish manure factory cry aloud that home is indeed the better place.

THE WITHY PEELERS AT WORK.

CAUGHT—FISHING ON THE NORFOLK BROADS.

VILLAGE ENTERTAINERS AT TAPLOW—THE ONE-MAN BAND.

ON THE ROAD AND IN THE LANE.

To many the country is a place of vast solitude where the plover wails over the deserted fields and the long ribbon of white road winds tenantless between the hedges or disappears over the skyline on the ridge. Of course, figures appear at long intervals, but they are as spectres which come from nowhere and thither vanish ; the labourer approaches a wayside cottage and is gone, the waggon is swallowed by the leafage of the lane, the fleeting motor-car dissolves into a cloud of dust.

Yet the most lonely road has its passers-by, men and women, young and old, rich and poor, worker and idler, sooner or later every class and every trade. The season and the hour bring their proper company so that the traveller can pick his companions and his mode of journeying, the idler observe the march of the calendar and the passage of time. On the road will be acquired not only the gossip of a county and news that does not get into the papers, but wisdom and philosophy that help to smooth the rough way of life, and knowledge that may add to learning, if not to income.

Spring adds largely to the numbers on the road ; it brings the pleasure seeker, the worker who prefers a country occupation during the finer months, and the migrant who earns a precarious living travelling from place to place. From his winter quarters emerges that vagabond philosopher the tramp, to beguile all who believe his plausible tale. Believe him when you have put him to the test, and if he withstand a severe cross-examination his reward may be regarded as well earned. His guises vary from the tatterdemalion rogue whose last desire is work to the tidy though threadbare wayfarer who carries all his belongings on his back and is anxious to turn an honest penny. Life in the open appeals to both, but the latter is the aristocrat of his class and besides being neat in appearance is perhaps the inventor of cunning contrivances that increase his comfort—sleeping bags, collapsible washpots, and so on. Not invariably workshy or fond of roving, the tramp is often a sailor bound for a port, an artisan going to work promised him, or a discharged workman making his way to an industrial centre and seeking odd jobs to help him on the road.

Many once familiar figures of the road seem to be passing along it for the last time as they grow fewer every year. The old fiddler who delighted the cottager is very seldom seen, and the robust " singer " of ballads must hide in the remote places. The acrobat, the performing bear, the one man band, and others, are also rare visitors. The organ grinder is not uncommon but few country children are acquainted with the hurdygurdy man and his monkey. Some of these, with the " Punch and Judy " show, belong to the motley crowd which at the beginning and end of summer used to travel along the roads leading to holiday resorts. Fewer made the return journey, as those who had prospered took the train, though some liked to make a regal progress spending their season's savings recklessly, and others worked a devious passage through village and townlet to add to their earnings. Times change, and these, though neither so numerous nor so representative of callings, may sometimes be met trudging along a wild upland road that is a short cut to their destination. The poor derelicts of humanity who seek to avoid the workhouse by appealing to the softer side of the prosperous are very familiar figures on the highway to the holiday town.

The migratory labourer who worked his regular round, hay time, harvest, and potato digging passes less often, but his place is taken in some districts by the pea and fruit pickers from the towns. Even the hop-pickers travel mostly by rail, but, occasionally, a family or small community may be seen on the road, pushing their belongings in perambulators or make-shift hand-carts. The once flourishing credit draper who drove in a light cart from village to village has been pushed out by the mail order system and cheap fares, but his brother the commercial gentleman is taking to the road again, astride a motor-cycle or travelling more comfortably in a roomy car with his samples.

However, the old users of the road have by no means all disappeared. In sparsely settled districts far from centres of population the pedlar finds plenty of customers although no longer the indispensable trader he was in a more simple age. He still sells the cheap finery and odds and ends beloved of the country damsel or useful to the housewife. As he approaches the hamlet he unslings his waterproof covered pack and sees the articles are well displayed before knocking at the first door.

Another old time itinerant regularly passes down the village street and stops at isolated cottage and farm, the vendor of the almanack. When he comes to a group of cottages he uplifts his voice in stentorian chant, proclaiming the virtues of the old " original prophetic almanack," its interpretations of dreams, its cures for the ills of the flesh, its garden hints, together with dark references to the terrible events foretold, an epitome of contents few can resist. Round his hat is fastened its gaudy cover or poster, slung across his back a pack or sack, and under one arm a sheaf of the publication which he sells as fast as he can hand them out. His hands appear to be fully occupied but occasionally he manages to waggle a bell which will attract attention when his chant may not.

In the country a bell is the never failing method of arousing attention, used alike by the crier, the newspaper boy—for nowadays the villages get their news regularly, brought by a boy on a cycle who finds a bell useful to announce his arrival and so save time in distribution—the ice-cream vendor, and the travelling ironmongers and crockery dealers whose carts are half hidden by their wares, brooms and brushes and the like. Unsuspected drawers yield cleansing materials and odd trifles for household use.

The itinerants are a numerous troop who call at cottage and farm. Some like the butcher and baker—the latter often met at cross-roads by children from isolated dwellings—have their customary rounds, but others appear at irregular intervals. Such are the knife grinders, chair and china menders, and umbrella repairers. Any of these may be seen at work on the village green or waste strip beside a hamlet, the more properous driving round in a cart, sometimes attended by wife or boy who acts as canvasser and assistant. The townsman is surprised when a fisherman met casually in a lane opens a basket and offers fish, but such pedlars may travel six or seven miles from the coast.

A cheery user of the road is the postman; perhaps his usually humorous outlook on life is largely the result of a clandestine perusal of the postcards that pass through his hands, for, though he denies the impeachment, sooner or later when turning a corner you will come on him glancing through the postcards and looking at the writing and post-marks on the letters. How otherwise could he tell such interesting tit-bits about absent relatives or the wanderings of the son from the " big house ? " And his questioning remark as he hands a letter over the gate shows plainly that if he does not know the answer he possesses a pretty good clue to it. He is acquainted with the secret romance and hidden tragedy of many a home, not necessarily gleaned from illicit examination of correspondence, and it may be his message that brings the maiden to the trysting place. He is usually obliging and ready to give a prosaic but urgent order to the grocer. Often he carries a whistle on which he blows a shrill blast when approaching a house off the road to announce his arrival in case the occupants have a parcel to send or desire to purchase stamps.

More dignified than cheery is the village constable with ponderous tread, whose burly figure has only to appear in the doorway of the inn to convert the most truculent to a belief in arbitration. The cyclist patrol represents the law upon the highway, having almost completely superseded the mounted policeman, not to the entire satisfaction of evil-doers. In his season and in the districts he frequents, the Breton onion seller is a well-known traveller, familiar with the lanes and foot-paths of the neighbourhood he visits annually. He invariably keeps to the high road, but when he leaves it and has to ask the way of strangers, he finds it difficult to understand or be understood.

All wanderers in the by-ways are acquainted with " Idle Jack," " Silly Sam," " Lonely Tom," or whatever be his local name, the squatter who lives in a tumble-down dwelling on the edge of a lonely wood or in the middle of a gorse surrounded fastness on the common. A lover of independence and the open air, disliked by the keepers, the despair of sanitary inspectors, often the butt of village children, he earns a living in a hundred different ways, and now and then is the suspected possessor of a well-filled stocking. Shy and often furtive, occasionally a cripple, he may peddle groundsel, watercress, and wild-fruits which he carries in a self-made basket. The acquisition of a box on wheels fills him with delight, and if he be shrewd and hard working, as are many of his kind, he ultimately becomes the owner of a pony and rough cart. Sometimes he is the son of a squatter like himself and inheriting a few worldly goods is mildly prosperous for he never lacks occupation, being ready to turn his hand to anything. Too fond of the inn, and an occasional poacher, perhaps, he is one of the characters of the road unlikely to survive much longer, for ever increasing regulations make his existence more precarious and he has to spend much of his time evading the authorities.

In spite of many disappearances the road has been awakened into life again by the motor-car. The great waggon long since ceased to make its regular trip of several days, taking away the produce of the country and bringing back other goods, but in its place has come the ponderous motor lorry. Only on very few roads does the coach make a gallant stand, for the war temporarily stopped many which have either remained in retirement or been replaced by motor omnibus. Twenty years or so ago the mail coach was quite common on cross-country routes ; some of the drivers and guards could regale passengers with many a romantic and exciting story of old days on the road, but where passengers were few the smaller mail cart supplanted them one by one, and now the motor

THE WAYSIDE HAMLET—SAINTBURY.

RURAL TRANSPORT—AN OLD-FASHIONED CARRIER'S CART.

COACHING—AN OLD INSTITUTION DESTINED TO BE MERELY A
PASTIME FOR THE WEALTHY.

LECHLADE—THE TRAVELLING STALLION.

A SKILLED HEDGER IS AN ARTIST AMONG CRAFTSMEN.

A NOTABLE COTSWOLD ROADSIDE FIGURE—PARISH CLERK,
MUSICIAN, AND GREEK SCHOLAR.

AN EVER-WELCOME CALLER.

THE MODERN REPRESENTATIVE OF THE
DEVON PACK-HORSE.

is ousting the horse their meagre passenger accommodation is disappearing altogether. Even the famous Minehead-Lynton coach has at last been ousted by the remorseless motor-bus from the road it traversed for so long, and now contents itself by making local excursions over new routes. Only the middle-aged traveller has experienced the discomfort and drawbacks of coaching in winter.

The carrier's cart lingers in many rural districts as a link with the past, but it, too, is changing in appearance if not in ways. Its hall-marks were a horse that knew the way and stoppages as well as the driver, a dusty tilt or equally dusty curtains, paint that had not been renewed for years beyond count, a cargo as varied as large, and passengers as loquacious as laden and portly. They drew up in the by-streets and inn yards of market towns, and those who watched them disgorge wondered how so many persons and so much merchandise could be squeezed into so small a vehicle. The carrier was a character worthy of his place in country literature. He was a perfect mine of information about the people, the happenings, and the legends of the neighbourhood, a merry man withal, and a useful one too, ready to collect debts or match the spinster's silk.

Such was the carrier's cart up to seven or eight years ago ; now it is very different, yet like all country institutions has many of the old characteristics. The horse is disappearing, and in places the " cart " has become a double-decked motor bus. It might be remarked that on shorter routes where passengers were many and parcels few the superannuated London bus often ended its days, announcements concerning Cheapside and the Strand being only half hidden by a coat of paint. The modern motor vehicle is a carrier's cart-bus hybrid; it picks up passengers at cross-roads and farm, its driver flirts with the girls at places where it stops for more than a few minutes, and it seems capable of accommodating all comers, for kindly policemen look the other way as folks must reach market, but it is somewhat alarming to see a bus careering down a steep winding hill with passengers sitting on the stairway.

However, the bus proper is confined to comparatively few routes and the motor " van " serves many markets. One type should help to solve the problem of rural communication because its cost and upkeep must be lower than that of the bus. It consists of an open lorry with stout roof for packages, well padded seats, and canvas sides with celluloid portholes that can be let down on rainy days, a most comfortable conveyance except when unduly crowded in wet weather. Some resemble a pantechnicon van lighted by small windows, with a tail board for produce and perhaps a folding seat on top in case of emergency, to be seen for instance in Hardy's Dorchester.   Sometimes a little motor van used for general purposes is pressed into service on market days when benches occupy part of the interior, the rest being devoted to packages and live-stock.   Fifteen miles in a tiny crowded van with pigs as companions can be enjoyed only by rural folk and those seeking adventure. Travelling in this way is a liberal education in country lore, though as conversation must be shouted to overcome the rattle exchanges are apt to get mixed.   Gossip of the district, rural wit of a more or less personal nature, original views on the topics of the day, discussions concerning prices, the state of the crops, signs and portents, and the new tenant at Eldergrove farm, are dovetailed into one another in a manner bewildering to those unaccustomed to it.

Deserted roads and lanes wake to life unexpectedly.   From a gate issue slowly one by one cows driven by a toddler whose chief office is to open and shut the gate, for the animals know what is expected of them ; from another comes the thirsty team on its way to the drinking pool, or maybe the passer-by must press closely into the hedge to allow an exasperated farmer to drive some pigs the way they should go.   Seldom are the grassy droveways thronged by cattle and sheep as in days gone by, but on the roads a passing drover or shepherd may tell of the long distance he has brought his charges, and in the vicinity of Dartmoor in early autumn mobs of cattle returning to distant farms are encountered.

Another animal of the highway is the travelling stallion, with glossy coat and arched neck, led by his groom.   Near the coasts of Cornwall and sometimes in North Devon a reminder of old time methods is met in the shape of a pony or more commonly a donkey with loaded panniers.   Not so very many years ago the pack-horse might still be found in a few remote places, but the day when it travelled in a train following a leader with jingling bells has gone for ever, being remembered only by narrow lanes lost amid tumbled hills, by pack-horse bridges, or by a set of crumbling crooks hidden in a loft.   In districts where the pillion lingered long after it had been disused elsewhere the upping stock or mounting block is a common feature.

To many the road and lane are the scenes of their daily toil.   Best known perhaps is the road-mender breaking the stones on the grassy margin near the hedge.   Apparently his chief desire in life is to know the time, and all who stop to chat after this simple request is satisfied find him versed in weather lore and often stocked with information concerning the life and habits of a past generation.   Some of these old chaps are also full of learning, as was the stone-breaker of Saintsbury in the illustration, parish clerk, Greek scholar, musician, an unusual but by no means unique example of accomplishments often possessed by those who follow humble occupations. On a shady spot near the high-road, down a leafy by-way, or in a hollow of the heath, is the camp of the travelling road-men ; beside the sheeted steam roller is the caravan in which they live, and when work is over for the day they may be observed over the open hatch titivating themselves for the evening visit to nearest town or village. Sometimes there is quite a company, three or four rollers and caravans laagered in an open sided square ; the crews sit round the fire on which they cook their suppers, listening to the strains of concertina or gramophone or merely smoking and gossiping.

One artist of the roadside would be sadly missed were he to disappear, the hedger. To cut a hedge is not a difficult task but to trim it so that it does not grow tall and spidery and full of holes quite another, and nothing delights the old hedger better than to show how his work should be done. Deftly he cuts and slashes and twists, and when he has done he leaves a green wall as level on top as if it had been planed and as close and velvety along the sides as a piece of deeply woven pile. If you doubt his skill walk a little further down the lane where is another hedge, thin, poor and full of gaps which must be patched with hurdles or bundles of brushwood to prevent the sheep from straying. The comrade of the hedger is the ditcher, an important worker if a shade less skilled. In hilly countries the ditches must be kept well cleaned or the water flows into the roadway, washing out the binding soil and scattering the stones. When the slope is slight the ditcher must know his craft or the water will gather in stagnant pools instead of draining away. Often hedging and ditching, being complementary occupations, are carried out by the same worker ; both jobs being piecework, they are often accomplished leisurely and interrupted by chats with passing acquaintances.

Gates are nowadays mostly made at the town saw mill and hung in place, a matter of a few minutes, so the gate-mender does not ply his trade by the roadside so often as he did when his working place was just within the wood. However, if necessary, an odd job carpenter removes a decayed gate-post and selecting a log from the stack under a lean-to shed in the farmyard roughly shapes it and drives it into place, using the old staple or hook for the latch unless rusted beyond repair. Here is a stonemason repairing a wall with much deliberation, there two or three building stone culverts in place of the wooden beams that for years have carried the lane over a drain, elsewhere a post and rail fence is being repaired or more probably replaced by wire. After autumn gales a gang is busy sawing up and removing the fallen elm that blocks the road.

In the by-ways a cottage wife plys her endless household duties, busy over the wash-tub, or preparing a midday meal. Some roadside occupations are carried on almost furtively, as is fuel gathering in times of stress ; two or three sally forth with pram or push-cart, in the bottom of which, often under the baby, is a hook. In an unfrequented lane where hedges are untrimmed likely stakes are hastily cut until the party have obtained as much as they can carry. The cotter possessing pony or donkey also sets out with a hook to cut the grass on the road's margin, his horse tethered to the hedge while he works; he either leaves the fodder to dry for a day or stuffs it at once into the sack that acts as saddle on the outward journey. As dusk is falling a lurcher with a rabbit in its mouth leaps through the hedge bordering the bridle path and makes off at top speed for his master hidden in a nook lower down.

Thus to thousands of country folk the road is part of their existence. Almost as soon as they are born they gaze upon its alternately dusty and muddy surface ; they play upon it in childhood, pass along it to their daily work, and lastly are carried over it in mournful procession to the graveyard by its side. " Up the road," " down the road " are expressions constantly on their lips. Before the sun has risen on a winter's morning they clatter by on their way to work, and as daylight wanes they return home in little groups, bass bags slung across shoulders, little firkins, perhaps, swinging from belt or hand, or a huge umbrella tightly clasped. They lollop down the lane with the peculiar gait of the countryman, smoking and discussing their work or some item of the day's news, and nod a cheery good-night to all who pass them by.

Along the road strolls the self-conscious youth with shiny face and huge button-hole, going a-courting in the next hamlet, the family party in their Sunday best proceeding on a state visit to relatives, and the ancient from the workhouse who likes to rest by the way with bandana handkerchief over his face as he takes a nap in the sun. Here is the roysterer rousing the echoes with his artificial mirth, there the erring son or daughter creeping back shamefacedly, anxious to reclaim their place under the old roof tree. Every place has its Lovers' Lane, of no importance whatever as a thoroughfare but rich in shady nook and overhung gateway, shrines devoted to Cupid and occupied by a couple of devotees who mostly worship in silence.

THE SUPER TRAMP—CARRYING HIS HOME ON HIS BACK.

THE TRAMP OF TRADITION AND FICTION.

BRETON ONION BOYS ARRIVE AT PLYMOUTH.

IN A DEVON LANE.

Poor must be the company if you feel lonely on the road except on the heath after dusk or on those misty days of autumn and winter when the moisture hangs low over the land and the solitary traveller is encompassed about by a filmy wall his eyes cannot pierce.

XI

TRADING IN THE COUNTRY.

TRADING in the country still largely remains a matter of primitive barter. Exchange is the order of the day, going by the name of contra accounts. It is an easy matter until settling day comes, say every six or twelve months, when one of the parties visits the other, spending the evening comparing accounts and settling values. Coals are weighed against joints of meat, dairy produce against groceries, a funeral against bread or feed for a horse, repairs executed by carpenter or builder against ordinary household commodities. Proceedings may be prolonged if unduly complicated, making a satisfactory settlement difficult to reach, and sometimes it leads to the estrangement of old friends and brings grist to the lawyer's mill. Yet it is not without its advantages, as among other things it enables the village shopkeeper with little capital to carry on and makes one without much money passing rich, but it is only in the remote and unsophisticated districts that it still flourishes to its fullest extent.

It is unwritten law that he who lives by the district must help the district to live in return, and that is why the new-comer who deals with the big London stores is so disliked. The local shopkeeper's point of view is really not so wrong, and, let it be whispered, he cannot be flouted with impunity, for the day may come when his goods are wanted, but he has only sufficient for his regular customers. Should crowded shelves and bulging cases cast doubt on this statement he answers testily when pressed " Buy your biscuits where you get your other things," a reply often given to a customer who usually favours a rival shop.

The wealthy are expected to turn a blind eye to illegitimate profits, and they are supposed to be above mere haggling or standing strictly by their rights, for what may be termed the sumptuary laws of the country come into force. The size and importance of a purchaser's house largely determine not only what price he must pay but what he shall buy. Even the itinerant hawker has a varied tariff, and the occupants of a house sandwiched between cottages are charged less than those who live in a smaller one among more fashionable company. To live in a large house yet possess plebeian tastes fills the village with horror. Says the butcher when asked for a joint he considers unfitting, "Oh, I have put that by for Mrs. Smith, the gardener's wife; surely you don't want that, only the cottagers eat *that*!" so much scorn being put into the concluding word that few have the strength of mind to persist. All such deeds of ill-fame become the talk of the place, and those guilty of them earn a reputation difficult to live down.

Shop-keeping in the country may be described as a hobby rather than a business; many traders cherish their goods like rare possessions from which they can be parted only by persuasion or guile. The shop, perhaps, is merely a side line, run more for the purpose of meeting all and sundry and hearing the latest gossip than for making money. Frequently a woman opens a shop to add to her husband's earnings; maybe, the "shop" is the front room of a cottage wherein are dispensed lollipops and gingerbeer; maybe a kitchen chair beside the door on which are displayed produce from the allotment, fruit, a jar of honey, poultry, crabs and lobsters in the fishing villages, or shells and polished pebbles in holiday districts.

It would be difficult to define many establishments whose proprietors describe themselves simply as "shop-keepers," sweets being common to all whatever the staple commodities may be. The saddler probably sells fruit and eggs; the butcher greenstuff and dairy produce; the grocer anything in the edible line except butcher's meat, with household materials and seeds; and almost every shop deals in proprietary articles in packet and tin. Some little shops can almost equal the vast stores for variety, but the sign or show-card is no index to the interior, being perhaps the legacy of a former tenant or merely displayed to attract.

Every wise housewife learns to know a shop as intimately as the proprietor, as he is sure to forget some of his wares. Sooner or later he may yearn for a change and then the draper becomes a grocer, the grocer a dealer in stationery and fancy goods, to quote extreme instances, but the transformation is usually gradual unless there be a change of premises when it may startle by its suddenness. It is also advisable to learn the signs which denote that an article has been sold out, such as reversing the cream and butter bowl; to ignore such cryptic announcements is to be received with scorn. Then when all this has been mastered the shop-keeper decides to retire from business and his window is draped with curtains, the shop becoming a sitting room. One day, as if he hankers after his old calling, the curtains are drawn coyly aside to exhibit some article of commerce, eggs and vegetables, fruit or a rabbit, or some still hoarded portion of the old stock.

In the market town certain classes of shop predominate, notably the ironmonger with his seasonable display of tools and implements, and the outfitter and bootmaker. The corn-chandler of the old prints is also there, in spring probably with a few chicks in his window to show the value of a poultry food, and the draper of a bygone day with rolls of material flanking the door, but they are becoming more rare as they blossom into the up-to-date establishment. Important are the numerous eating places, for your market-goer is a good trencherman ; they are many and various, from the hotels and inns where the market ordinary is provided, and the confectioner-restaurant with substantial dinners and equally substantial teas, to the small eating-house in a side street which caters for the carters and drovers. Often a modest looking shop displays merely a joint of beef and some vegetables, but such a place cannot be judged by its appearance, as it may be the most popular of all, with large rooms, though hardly big enough during the dinner hour.

The greengrocer and dairy are almost confined to the market town for they are rare in the country except in holiday districts. A local market-gardener usually provides fruit and greenstuff, and the shop-keeper with cows sells dairy produce as a side line. The farmers have their milk rounds usually carried on by a small boy or even a girl. The jangle of his cans wakes the sluggard, and in a basket he carries eggs and rolls of butter wrapped in spotless cloths. It will be a sad day if legislation banishes his familiar figure ; it will bring little advantage to the boy as the work is not arduous and he gets a good breakfast in addition to the small weekly wage, besides obtaining some insight into the work he will take up in after-life.

Shopping in the country can be exasperating, and it certainly teaches resignation and contentment with little. The hawker who brings round potatoes and vegetables may arrive or he may not ; he has gone out rabbit shooting or his horse has fallen lame. The shop-keeper may consider it a fine afternoon for a stroll so he shuts up, and at hay time and harvest all the village shops may be closed or in the hands of the very aged or very young, when it is inadvisable to buy. An outing or a visit to a relative may also cause a cessation of business, but every-one is supposed to know and arrange their wants accordingly.

Book-keeping is a mysterious art better kept at arm's length, and a simple calculation the monopoly of the learned, so if the joint does not weigh an even half pound, a sausage or an odd scrap makes up the round weight. Local weights and measures together with methods of sale are also mystifying, for a district is often a law unto itself ; here everything is sold by the pound, there by the gallon, elsewhere by the number, and there are local weights unknown to the majority.

Yet with all its exasperations shopping in the country is not without its enjoyments. At its best it is a friendly transaction carried on at leisure, and the customer is invited to see the new baby or shown a family treasure. Some of the shops are delightful, ancient buildings with bellying fronts containing windows of many panes, sagging roofs of aged slates mottled with lichen and moss, old carving and plaster work, with odd chambers and little nooks inside. Others are pictures of beauty overgrown with rambler roses, honeysuckle, and jasmine draping the window in which jars of sweets mingle with string and apples, and picture postcards peep from behind packets of grocery and tobacco.

If country folk enjoy shopping, as it affords such an excellent excuse for a gossip, they fairly revel in sales. A sale at a cottage will set a large village agog with excitement if it be one to which entrance has been restricted to the few, a rare occurrence as it is seldom the precise contents are not known to all. Usually the sale takes place in the afternoon, the morning being set aside for the preliminary view when everyone seizes the opportunity of learning the style in which a neighbour lives. Numbers attend the sale more for the fun of the thing than a desire to buy, especially when the auctioneer is reputed to be a wag, and most of them know how to bring forth a laugh which helps to increase prices. Treasures may still be picked up at marvellous bargains, especially in out-of-the-way places, but, on the other hand, it is surprising what sums are bid for ordinary articles much in demand, and when popular neighbours fall on evil times. After a sale a curious procession wends its way up the village street, women sharing the burden of arm-chairs and baths, crockery or a bedstead on a wheelbarrow, a piano laboriously carried by two men who put it down every few yards, or a studious young man carrying an armful of assorted books.

Farm sales, numerous at Michaelmas, provide an excellent opportunity for meeting acquaintances. The men bring their women-folk who regard it as a fitting occasion to compare babies, and most are dressed in their best. A stream of traps and horsemen with a few motors passes up the lane to the farm about which the horses are tethered, especially along the hedge of the field in which implements and waggons are arranged, each with the number of the lot chalked upon it. Beer or cider is handed out to all and sundry, with a nip of something stronger for special friends, before the men drift off to inspect stock and implements or sample grain to be sold.   After closely examining furniture and utensils the women gather in the kitchen or large parlour to criticise freely or discuss intimate family details, odd scraps of conversation giving a clue to the cause of the sale.   The family retire to a smaller room, which must be constantly defended against blundering intruders who retreat covered with confusion.

Would-be purchasers keep as near the auctioneer as they can or they may be crowded out of the small room in which their desired bargain is sold, and bidding from the stairs may cause inconvenient mistakes.   At farm sales in districts difficult of access furniture sometimes fetches more than its original price as buying on the spot saves transport.   After the sale the auctioneer repairs to a table in a corner to receive the money, and one by one the buyers take their goods and depart, loading traps beyond normal capacity, but heavier articles are fetched by cart next day.   Sometimes a sale is in the nature of a farewell to friends, for the farmer is leaving the neighbourhood or, having failed, must seek his fortunes anew elsewhere.

Important, too, are the routine sales, the seasonal auctions of grazing land, corn in ground, orchards and fruit, and stock, usually held at an inn centrally situated.   Sometimes large sales of stock take place on comparatively small farms, herds and flocks being sent from the surrounding neighbourhood.   The lanes and paths converging on the scene are thronged with all classes, a gentleman or two, farmers, graziers, butchers, and labourers wanting a few pigs, to whom are added excited bullocks and frightened sheep, not to mention barking dogs.   In early autumn sales of wood for fuel take place on big estates, and sometimes a tradesman giving up business sells horse and trap and implements of his trade in village square or any open piece of ground.   Other horses and vehicles are often included on such occasions, making the sale a large one.   The annual jumble sale is another popular event which leads to weird scenes when purchases are carried home ; afterwards familiar garments appear in new guises, but the modern cottage woman is not quite so fond of cast off finery as she was, she prefers new !

Shopping is a daily commonplace largely confined to the women, and sales an enjoyable interlude, but market day is the proper time to transact business of any sort, from selling a pound of butter to settling a large deal in land.   Markets existed before shops, and as shops have waxed markets have waned, though many flourish as they did when first established several centuries ago ;   their commodities may have changed and methods of sale altered, but they remain indispensable to the life of an agricultural district. They are attended by country folk from miles around, and often the retailers in towns of greater modern importance depend on them for produce. A market brings business to a place which would otherwise stagnate, for outsiders with produce to sell have their own wants to satisfy which benefits the shops. The infinite variety of a good market is amazing ; farm produce of course, ironmongery and implements, ready-made clothing, stationery, crockery, toys, sweets and confectionery, and the wares of the marine store dealer, little seems missing.

Over it all is an air of liveliness and good fellowship which are irresistible, and an occasional quarrel or even fight being of serious moment only to the participants and a matter for conversation. Round the stalls and pens circulate the news of half a county, and a market would serve its purpose if it did little more than keep those dwelling remote in touch with their fellows. It is a reception room and rendezvous where all can meet without carefully arranged appointments, for if the market-place be drawn blank visits to certain snug hostelries will almost certainly discover the wanted friend. Except in the afternoon when they shop on their own account, the women will be found in the market hall—pannier market as it is usually termed in the south-west, a survival of the day when lighter goods were carried in panniers—sitting in their accustomed place behind a rampart of fruit and vegetables, bowls of cream, and butter.

In the afternoon and early evening comes the bustle of departure. Distracted carriers seek to round up belated passengers engaged in final gossips ; shouting drovers and barking dogs keep herds of frightened stock from straying ; horses and traps pop out of inn yards like shots from a gun as ostlers love to send them off with a resounding smack ; here are two farmers engaged in a last deal ; there a Shire is still being put through his paces in the fast thinning market crowd, a sure sign that price is too high or some weakness present.

Open-air markets, with booths arranged along the kerbstone are common in the smaller towns, and on remote hill-sides are "markets" where sequestered farmer and cottager meet traders from the nearest town, or once a month dealers in crockery and ironware gather at a convenient centre, a little strip of level ground in a sheltered spot. Important are the special markets and fairs ; the numerous horse and pony fairs ; the Christmas cattle markets when the main street of a small town is given up entirely to stalls and pens ; the poultry markets ; and the cheese fairs of Yeovil and Melton Mowbray.

One is tempted to quote old market rules and regulations against profiteering in all its forms which are as old as the market itself. So that all might be served alike some ordained that no buying or selling might take place before the ringing of the market bell, others that no outsider could buy until local requirements had been satisfied. Market prices ruled over large areas as they do to-day to some extent, and the authorities of a few market towns levy toll on all goods sold within their jurisdiction by anyone dwelling outside. "Market clothes" and "market merry" are expressions whose full significance is almost forgotten where market day is no longer an important event of the week.

Isolated farms and cottages are served chiefly by the itinerants, a numerous band. The village butcher has his round, his cart tail being counter and chopping block, sometimes taking back eggs and farm produce to sell at his shop; the grocer whose cart is covered with advertisements and hung with household requisites; the oilman who has a carefully mapped round for every day in the week; the leather seller who supplies the cobbler and those who mend their own boots; and the tin and china-ware trader and travelling ironmonger who keep to the villages and larger hamlets, seldom turning off their direct route to visit scattered groups of houses. Near the coast the fish hawker is a regular visitor to town and village, especially welcomed in the herring and mackerel seasons, a trader whose special cry is his own personal trade-mark although each declares his fish are the freshest. Where catches are landed on the open beach he sometimes sleeps out under hedge or boat so that he can be up and away to catch the cream of the market.

Eagerly looked for in summer by old and young alike is the ice-cream vendor whose gaudy barrow visits the most secluded hamlet and halts at quiet cross-roads; in winter he may go in for hot potatoes or chips, but his round is a short one, being confined to the evening. The tinker and chair-mender never lack custom though they seldom run to a horse and cart.

Cheapjacks are perhaps not so numerous as they were, and the quack doctor and dentist no longer such familiar figures, probably crowded out by more ambitious representatives and to some extent by the State Panel doctor. Some were of good repute who worked a neighbourhood year after year, announcing their coming in the local press, but others took care not to return until well forgotten. Among the "doctors" are qualified pharmacists possessing "the gift of the gab" who find they can earn more in the market-place than as assistants behind the counter.

EAST HENDRED. BERKSHIRE.
" A thriving and well-accustomed village shop."

"The Tinker, notwithstanding his vagrant habits, is sometimes a man of substance."

THE MILLER'S CART—DORSET.

HORSE FAIR, BAMPTON, OXON.

HIRING FAIR AT BURFORD, OXON.

THE STOCK SALE AT "OFFIELDS"—CHURCHSTOW.

A MARKET SEVEN HUNDRED YEARS OLD.
Market Drayton obtained market rights in 1245.

Some spend the summer travelling and winter in the shop, some travel all the time, but a few do so well or have such a reputation they remain at home during the inclement months, living on their summer earnings and the profits brought by remedies ordered by post. Most declare business is not what it was fifteen or twenty years ago, and some aver that their pills are similar to, but fresher than, those of the country chemist who has not nearly so large or quick a turnover. The inn parlour or empty shop is their consulting room, the market square the scene of their " cheapjackery," and those jealous of their reputation advise a visit to the doctor when necessary.

An important member of the trading community is the carrier, who will fetch anything from a large case to a packet of pins, and is usually ready to match a pattern if required ; but whether his modern brother of the steering wheel will have time for such good nature remains to be seen. A mysterious occupation that might almost be termed anonymous is that of the Sunday newsagent. Most of the Sunday journals have their own agent who meets the carrier at some unknown spot, and the strict church-goer receives a shock to learn by chance that the much-respected verger distributes the most lurid of them. The cottager would be lost without the journal which supplies him with such news as he likes—battle, murder, sudden death, and the sins of society—and the " literature " he desires, moulds his politics, and gives him advice on anything from tending his garden to his legal standing in the dispute with his neighbour.

Many and various are the trading methods adopted by the country-man according to the requirements and customs of his locality. His description of himself may be a poor index to his activities : the corn dealer may not only deal in corn and feeding stuffs, fertilizers, and so forth, but be miller, farmer, insurance agent, coal merchant, and estate agent. The builder is usually as versatile ; and the coal merchant and carrier often combine several trades, while the inn-keeper is usually also a farmer. It is difficult to understand how some of the superior cottagers earn a living ; they never seem particularly busy, though not averse from engaging in many odd jobs. On inquiry it will be found they have a horse and trap on hire, they buy calves at one market and sell at a profit at another, they take droves of horses to distant fairs, they rear pigs, or do a little butchering at times. Industrious labourers add to their wages by cobbling shoes, or gardening, or by engaging in other tasks in which they possess skill.

Neglect to observe the ancient rites inseparable from trading in the country is a sign of eccentricity if not a minor crime. Hurry or instant decision may almost denote ill-breeding ; and to step in while another is " thinking it over " is something like sharp practice, although there are places where it may be commended as business aptitude. However, sharp practice is no more absent from trading in the country than in the towns, and many an innocent looking rustic can give points to the smartest dealer in the city, but on the whole country trading is a leisurely and kindly business. Nevertheless, few are sufficiently sentimental to protect the fool from his folly, and the person in a superior position is frequently favoured at the expense of his less prosperous neighbour—although he must always pay for his privileges and must not expect to get it both ways—partly because it is a survival of the old feudal tradition, and partly because the countryman has had such a struggle to live, he has learnt to make money where and how he can.

THE ICE-CREAM MAN—ALWAYS WELCOME ON HIS ROUNDS.

A CORNER OF THE "PANNIER" MARKET.

THE PIG ROAST AT BIDFORD MOP.

ASSEMBLING FOR THE CLUB WALK.

XII.

THE COUNTRYMAN IN HIS LEISURE.

SOME would have it that English country life is dull and dreary, that fun and frolic departed with that period known as the time of Merrie England, when the calendar was studded with festivals, saints' days and holy days, when there was tilting on the green and merry-making in the inn. Folk were more primitive then and more easily amused, without the leisure they enjoy to-day, so the Church ordained these feasts as pleasant interludes amidst incessant toil. It is often contended that Puritanism sounded the knell of Merrie England, but although it had great influence and now and again managed to impose its own dour solemnity on an unwilling people, it never really succeeded in preventing the countryman enjoying his traditional diversions. The gradual change in habits, the slow development of taste, and the greater ease of intercourse have had most to do with the passing of the old ways.

Village life is by no means a dull round of toil broken by duller leisure ; dullness exists, no doubt more frequently than it should, but an absence of artificial pleasure does not necessarily mean boredom, while the heavy bucolic type is almost incapable of gaiety although inclined to joviality. Of old, village amusements were communal like its life ; people provided their own entertainment, which to-day is one of the differences between town and country. The village thoroughly enjoyed its miracle plays acted by itself, and now some home-made entertainment in the schoolroom gives more real pleasure than that provided by the best professionals. It smacks of the district and is studded with local phrases, allusions, and jokes, all the more popular from frequent use.

Being innately conservative the countryman spends his leisure largely according to habit until it becomes a sort of ritual ; variety he scorns and is happiest when following the same programme day after day. Should it be altered be sure that haytime or harvest, illness or the fair has merely caused it to be changed for the moment. Unless the weather be unpropitious the younger people enjoy their promenade up and down street in groups of youths and maidens, exchanging rustic wit as they pass, but before the evening has grown old the groups mingle and break up into smaller and mixed parties or couples, for in this manner country courtships often begin. The older folk stand or sit by their doors, the women knitting or gossiping, the men smoking in contemplative silence except when they hail a passing comrade ; as the dusk gathers the children are called in from play, but the men withdraw only when the lamp is lit and smoke curling from chimneys announces the preparation of supper.

The station, especially one overlooked from bank or bridge, attracts those who like to see the evening train come in, and when the village threshold is a bridge over a stream, there the men foregather, leaning carelessly against the parapet and discussing all things—from their allot-ment to the latest murder. Such a bridge is second in popularity only to the inn whose disappearance would leave a void impossible to fill ; it is a club where the labourer can feel at home and can talk over those matters which affect him closely. In summer, he sits on the benches in the open, in winter, in the settle round the blazing fire, the work-aday world shut out by the red curtains drawn across the windows. Generally it is a staid and eminently sober gathering, presided over by the inn-keeper, but truth compels mention of dirty little ale-houses of ill-repute where quarrels are frequent and the landlord is as rude as his customers are rough. Formerly the labourer enjoyed harmless amuse-ments at his inn, throwing darts at a target, playing a kind of wall quoits in which a ring is cast upon hooks, and so on, but such amusements are frowned on if made merely an accessory to surreptitious gambling.

Another time-honoured habit is the Sunday evening stroll. Every place has its recognised parade, a pleasant road beside a river, a by-way across a down or common, a lane, or footpath over meadows, with the established route out and home. Here the tradesmen promenade in slow and stately fashion, the men in sober black, the women in rustling silk or satin, often of ancient fashion ; here, too, are the labourer and his wife with their numerous brood running round them, peeping into the hedge and gathering wild-flowers, withal indulging in no unseemly romping unfitted to the Sabbath. Early closing day is also largely devoted to a stroll ; in summer, it affords an excuse for a picnic or a trip to the seaside ; in winter, an opportunity for visiting and tea-parties ; and it is the appointed day for entertainments in the local hall.

The countryman is gregarious and likes a crowd ; a single day's outing with his friends is worth a week's holiday by himself. An odd day off is probably passed in "busman's" fashion, doing odd jobs about the home or allotment or walking the lanes to watch others at work.

Perhaps to mark the occasion he directs his steps towards the inn, but as a rule the solitude of his own society, punctuated by a brief visit from a passing carter and a broken conversation as the landlord pops in and out, soon palls, so out he goes to gossip with whoever he may meet.

Such an odd holiday may be utilized for the annual visit to relatives, that solemn undertaking which requires such careful forethought and preparation, anticipated with misgiving lest Mary or Tommy disgrace the family, and is completed with relief. A family party setting out on a state visit may be recognized by their preoccupied air and careworn expression, the frequent admonitions to the children to behave quietly and keep themselves tidy, and perhaps the parcels they carry contain gifts to disarm criticism. In these prosperous days it may even be made in style by motor car.

If the family visit cannot be considered in the light of a real holiday, a day's outing with a party is something to anticipate with pleasure, alloyed with doubts about the weather, and to reflect on and talk about for months. Some join clubs and societies merely for the annual outing, but as relatives and friends may usually be invited, those with extensive family connections—and who has not in village and little town !—are seldom at loss for an excursion. Lamentations only arise in the unlikely event of two parties on the same day. Early rising is necessary not only because the start is early but to choose attire suitable to the weather and pack the eatables, both matters of importance on such occasions. Motor char-a-bancs encircle one county or cross two while horse brakes make shorter journeys to favourite spots where most of the time is devoted to hilarious games interspersed with dinner and tea. Bellringers may spend an arduous day ringing the changes in the churches they pass, which may shorten the length of their journey but does not curtail enjoyment or hinder refreshment.

Outings benefit the few, but a local fête sets a whole neighbourhood talking. From the preliminary puff to the publication of the balance-sheet it affords material for conversation. Unpropitious weather cannot altogether mar its glories, and the more closely it conforms to the traditional pattern of the district the better is it liked. Such fêtes vary little, though minor details and events reflect local idiosyncrasies ; " Aunt Sallies " in the form of the most unpopular man of the hour, the donkey that moves head and tail when the bull's eye is hit with a ball, throwing hoops for prizes, guessing the weight of pigs, and so on. Grinning through a horse collar or climbing a greasy pole has a place

sometimes, but we live in a sedate age when " A Match at eating Hot
Hasty Pudding by Boys " or "A Match at drinking Hot Tea by elderly
Ladies " is unlikely to get beyond suggestion by the most irresponsible.
A whist drive is another popular feature although it means missing much
of the fun of the fair.

Surrounded by knots of supporters, the local athletes of reputation,
more or less suitably attired, stand aloof, either wearing expressions of
gloomy determination or volubly explaining why they failed to win in
spite of noisy encouragement. A young woman with determined mouth
and chin usually wins the ladies' race, because the others let laughter
overcome pace, and as a grass field makes hard going the ancient who
manages to complete the course is the most likely victor of the veteran's
walk. The freak contests are always warmly welcomed, and it is a
good fête that includes a comic cricket match or pony races. Tea is
not the least popular feature, but supplies have an unfortunate habit
of giving out before all are satisfied, and the local band comports itself
with the dignity of a trained orchestra. No one enjoys it more than
the members of the managing committee whose pleasure is long drawn
out. Their part is not confined to fussy importance on the great day,
but includes the talkative preliminary meetings and the final wind up,
when congratulations are mutual, perchance with a dash of happy
criticism because a much discussed innovation has fallen flat.

Closely allied to the annual fête, but of far more moment, is the
celebration of some event of importance, a Jubilee, a Coronation, a Peace
Day, or a local Centenary. That is indeed an affair of great magnitude
which lasts the whole day and embraces all in the parish, proceedings
usually opening with a service in the Church, and ending with a bonfire
and fireworks, and of course the inevitable dance. Streets are gaily
decorated, and the main items are the procession in costume, in which
village club and Scouts have their place, with prizes for the best dressed,
sports in the afternoon and presents for the children, perhaps tea or even
dinner in the street, dancing being started on the fair ground or some
spacious lawn and continued in the local hall. Villagers dearly love a
procession headed by a band, and the annual carnival, often on Bonfire
Day, is an institution in many places, its ostensible purpose being to
raise funds for hospital or charity. Modern methods of insurance have
largely killed the benefit club, so the carnival with its cavaliers and
Red Indians, to say nothing of its waggons bearing time-honoured
tableaux or new ones racy of local events, is more popular than ever.

The old men regret the passing of the Club Walk, when, headed by
a flag and band, they marched to church carrying wands or poles, those

ALL THE JOLLY FUN.

" Where village statesmen talk'd with looks profound,
And news much older than their ale went round."

THE INN PARLOUR—OVERCOTE FERRY, ST. IVES, HUNTINGDONSHIRE.

A man who relies on his skill as wildfowler and fisherman.
THE NORFOLK BROADS.

A WINNER AT THE CART-HORSE PARADE.

of the officers topped with emblems, afterwards dining at the inn and holding a sports meeting in the afternoon. Full of interest is the history of these benefit clubs, the ancestors of the friendly societies, which in the opinion of some originated in the church or trade guilds. As the rules of some bound their members to " live and behave themselves on all occasions religiously, honestly, and soberly " and to be at their " respective places of Divine worship every Sunday," the Church had probably something to do with their beginnings, and the fact that they meet in the local inn does not weaken the argument. In populous places members of the big friendly societies may parade annually, the Ancient Order of Foresters in appropriate costume accompanied perhaps by a waggon with Robin Hood and his company, or the Loyal Order of Shepherds with each member carrying a crook, but nowadays the business side receives more attention than the spectacular.

The Revel, Wake, or Feast still survives to some extent, notably in Cornwall, Devon, and parts of the north. Beginning with devotions it ends in mirth, and formerly was often marked by the preparation of special dishes, besides being regarded as the day above all others for family reunions. Sometimes it coincided with the fair, and a few stalls and a merry-go-round recall the old times to memory. Most fairs are only ghosts of their former importance, and it is difficult to remember they were originally wholesale markets held for purely business purposes, the amusements gaining prominence as the trading declined.

Unenticing as is the average modern fair its story helps us to understand bygone ways, and the old ceremonies connected with it may be observed. At Honiton, Devon, the fair is opened by proclamation of the town crier and a gilded glove is displayed, and at Seamer, Northumberland, it is cried, and the charter granted by Richard II to the Earl of Northumberland read in three places by the head-keeper, accompanied by a cavalcade of tenants. Peterborough rejoices in its " sausage luncheon " after the fair has been proclaimed on the bridge ; Stratford Mop is famous for the roasting of an ox, Bidford of a pig, and elsewhere similar or other customs still obtain. Sherborne has its Pack Monday fair, supposed to have originated at the completion of the Abbey, and does not the cuckoo make its first appearance at Heathfield fair !

Business and pleasure are still combined at a few fairs, chiefly the horse and cattle fairs and the few hiring fairs that survive. However, attendance at the latter is small in comparison with the past, but where the custom continues the labourer dressed in his best seeks a master—the carter carrying a whip or with a knot of whipcord in hat or buttonhole, the shepherd disporting a tuft of wool, to which may be added a bunch of ribbons when a bargain has been struck and the shilling "earnest money" changed hands. Now new methods are supplanting them in the north where they have held their own, and in the south that at High Wycombe will soon be only a memory along with Newbury, Abingdon and Burford, all well attended almost up to yesteryear.

Even the pleasure fair is declining although the merry-go-round, with prancing horses supplanted by strange beasts and dummy motor-cars, is as popular as ever, and the swings and shooting galleries, the aerial railways and other introductions, together with all the rest of the "jolly fun" never lack patrons. The panorama had almost disappeared before the cinema killed it, the wax-work show is uncommon, and the booth theatre and penny gaff unknown to the present generation. The travelling circus, whose arrival played such sad havoc with school attendance, has lost much of its spectacular glories and seldom possesses mangy lions to arouse the village with their roarings.

Perhaps the countryman enjoys his leisure most when sport enters into it. Habits may change but nothing seems to dull the edge of sport, though some counties may be more sport loving than others and more devoted to certain pastimes. The local race meeting is an excuse for shutting up shop, the jumping field is the most popular feature of the agricultural show, and men hang about cross-ways to ask how the first day's shooting has gone or who has won the cricket match. The apparently deserted hamlet becomes populous when the horn sounds on a winter's morn, and all who can snatch a few minutes to see the meet on the green or watch the otter hounds start off up stream. The cricket match may often be responsible for delayed messages and the carter's late arrival. Sport brings all classes together on a common level, cements good fellowship and encourages kindly feeling ; the village where squire's son and carpenter, parson and inn-keeper, are regular members of the cricket team is a better place to live in than one which does not support such a commonwealth of amusement.

Rare is it to find a district that does not point proudly to its champion, a cunning fisherman who has haunted the local stream since boyhood, a wild fowler of the marshes and fens belonging to a dying race, a keen runner to hounds who never misses a meet near his home and knows the country so well that he is in at the death as often as the

THE WATER JUMP.

THE HOUND SHOW.

COLLECTING THE BAG.

RURAL SPORTS MEETING—THE START FOR THE 100 YARDS.

THE PLOUGHING MATCH—EAST ANGLIA.

THE MEET AT THE CROSS ROADS—DORSET.

"The chatter, the bustle, and the merriment of the Fair."

stranger mounted on the best of hunters. In one place he is the black-smith renowned far and wide as a mighty hitter or a bowler who has been the undoing of many a crack county player ; in another a plough-boy runner, the acknowledged champion of the district, or the proud owner of a pony which is the local pride at every Galloway meeting in the neighbourhood.

Local sports and pastimes of which one seldom hears have not been driven from the field by those more generally popular. Perhaps the war has put an end to hawking on Salisbury Plain, but Cornwall still enjoys hurling, and wrestling has not died out there or in the dales of the north-west. On the Somerset shore the seeker after the unique may sample the excitements of hunting eels with dogs, a furious, splashy pastime ; fell running and hound trailing belong to Cumberland ; Yorkshire has evolved the simple pastime of billets that demands much skill, and arrow throwing ; and Sussex maidens have long been famous for their prowess at stool-ball. Camping, that battle in miniature, has vanished from the fields of East Anglia, and Cornishmen no longer hunt the seal in his lair, a sport requiring great courage and never failing resource. Our sports are not so rough and brutal as they were, although knowing ones can sometimes take you to a cock-fighting main, and unscientific bouts at fisticuffs in the old style are not quite unknown.

The farmer enjoys his day's rabbiting as much as does the land-owner his sport with the pheasants, and the labourer who goes forth surreptitiously with ferret and terrier is often actuated as much by his love of sport as desire for gain. The popular athletic meeting attracts more people and is more orderly than Dover's Games, and the football craze is extending to the hamlet-village which competes in the local league. For quiet enjoyment broken by periods of excitement what can excel the regatta on the placid waters of an inland stream or in a winding estuary sheltered by encircling hills ? A picnic no less than a sports meeting as are so many rural festivals, be they the race meeting, the opening of stag hunting on Exmoor, or the point-to-point. And what more typical of quiet country life than lawn tennis in an old-world garden or on the lawn behind the mansion ? A bowling green may not be the usual adjunct to the country house as in times when pastimes were fewer, but the ancient and restful game is much favoured by the shop-keepers and elderly men of the smaller towns.

The country dull forsooth ! Why scores of events provide an excuse for a holiday. At a ploughing match the eyes of the hamlet follow the furrow of its champion and if he go off the line for a hair's breadth its sigh whistles like the wind. Neither the strident note of steam organ nor the popping of guns in the shooting booth pitched on the adjacent strip of waste ground can attract the backers of Young Will, the hedger, until he has completed his task more or less to his liking. None who have walked a puppy can miss the hound show or puppy walk, and how can the farmer and his men be expected to work when the agricultural show is taking place ? Suppose they were absent when their own carter led away Dappled Beauty with the winning rosette putting the finishing touch to its shining harness and tinkling ornaments! If you would see nonchalance struggling with disappointment—or anger openly displayed—watch the farmer whose bull has not been placed or the shepherd whose dog has failed him at a critical moment of the trials. Allotment holders, at the least, tremble with excitement when they go to see the decisions at the horticultural show. The judges may cause disappointment and heartburning, but are not so likely to raise the thunder clouds of disapprobation as the awards in the baby class.

Scores of local celebrations do not find their way into general calendars, cherry feasts, fishing feasts, and the rest, which help to fill the countryman's leisure, but for right down excitement you cannot beat a hotly contested election. Candidates in be-ribboned motors or coaches are received by cheers and boos in hamlet and at cross-ways, and workers in the fields almost come to blows over the merits of rival programmes which they hardly understand. The improving lecture is one method of filling an evening that rural folk do not love ; a lecture on cottage gardening seldom draws the labourer, and the women laugh at the mere suggestion that they have anything to learn about house-wifery.

In winter darkness comes early, and though more time is spent in sleep the long evenings are not so very dull. The inn always has its occupants, but the farmer is more of a home bird than he used to be and is no longer seen in his accustomed corner enjoying a quiet rubber with his cronies. The whist drive is the most popular winter pastime in small town and large village, and outlying hamlets hold them at intervals. These usually conclude with what has been termed " a little swing round," for dancing does not lose its popularity although the old country dances are almost forgotten, their place being taken by the latest thing taught by a mistress from the town. No festivity is complete without its concluding dance, and nowadays every place, however tiny, which boasts a hall, has its fortnightly dance to which people flock from all the neighbourhood. The village fiddler who knew the old tunes and ballads is nearly extinct, but two or three musicians with a reputation, though they may play entirely by ear, are always ready to oblige.

Science has come to the aid of those who dwell remote and apart ; when the wind howls outside and the rain beats upon the window pane or the country lies under a mantle of white they draw their chairs to the fire and turn on the gramophone, of whose strident notes they never grow weary. If the town be near, friends or neighbours make up a party to walk in to the cinema perhaps once a week, and at Christmas, occasional char-a-bancs excursions are run to the cities for the pantomime. Another achievement of science, which as it develops will tend more and more to contribute to the popular entertainment of country folk, is that of broadcasting by wireless-telephony. This marvellous means of communication, whereby the dwellers in farmhouse or humble cottage may remain at home and " listen-in " to the important news of the hour, to speeches as they are delivered, and to concerts, will do much to kill the monotony of the long winter evenings —especially in the more isolated spots of the countryside. Thus the amusements of town and country meet, and though their pleasures may not usually be the same, it is not necessary to lament unduly the dull lot of those who do not dwell amidst a populous community.

XIII

OLD-TIME CUSTOMS AND FOLKLORE.

CRITICS of rural England often declare that the countryman is too much the slave of ancient custom, that work and play are largely governed by the habits of generations but slightly modified by the leven of new discovery and changing circumstance. Yet what else could be expected? Only within comparatively recent times has the country been brought into close touch with the outside world, and agriculture must of necessity develop slowly. Everyone is influenced more or less by his work and surroundings, and thus it comes about that the villager is by nature averse from change and clings lovingly to the old ways, old methods, old beliefs.

The common customs that touch on every side of human activity, and date from all ages, differ almost from place to place. To classify them is difficult, to describe them in detail impossible, but taken together they form patterns in the mosaic which represents the life of England's country-side, its work and play, its sorrows and joys. Some have been described so often that they are familiar to all, and others are almost unknown beyond the confines of the district in which they are observed.

One may call country people ignorant and superstitious, but after all they have experience as a guide and more often than not some warrant for their beliefs. These beliefs are usually the survival of the theories and methods of long ago when the learned of the time dwelt in the monasteries scattered up and down the land. Others are older still, going back to days almost before the dawn of history; in these, ignorant superstition is, perhaps, mingled with a modicum of scientific fact and sometimes tinged by the poetical mysticism often found in the simple. The most fantastic may owe their origin to some rustic philosopher trying to puzzle out a mystery for himself, or represent the primitive use of a natural law stumbled on by chance. Many of the old saws and sayings and remedies can be traced to their source in one of these. Disentangle the sounds of a charm which seems gibberish and there emerges an artless prayer which brought relief and hope to all of simple faith, really a variant of modern fashionable faith-healing.

THE HORN BLOWER OF BAINBRIDGE, YORKSHIRE.

AFTER SCHOOL: MARBLES—WIRKSWORTH, DERBYSHIRE.

MAY DAY FESTIVITIES,
CHIPPING CAMPDEN.

"Round the Maypole, trit-trit-trot!
See what a Maypole we have got;
    Fine and gay,
    Trip away,
Happy is our new May Day."

CHARACTERS IN THE HORN DANCE. ABBOTS BROMLEY.

CHILDREN DANCING ROUND THE BAAL FIRE AT WHALTON,
NORTHUMBERLAND.

READING PSALMS AT THE COFFIN WELL, TISSINGTON WELL DRESSING.

However, a veneer of education has almost destroyed what might be termed the primitive culture of the peasant ; ancient customs are dying out and old beliefs losing their hold, but the survivals are many. From the cradle to the grave some part of the country celebrates the most important events of life or prescribes the rites that shall accompany them ; largely restricted to the nooks and corners though they are, all really intimate with rural folk marvel at their number and variety. It may be the proper person in whose arms a newly born infant must first be placed, or ensuring it shall rise in the world by carrying it upstairs before it is carried down. Some mothers give a grateful smile if on presenting the latest arrival a silver coin for luck be pressed into its tiny hand. In the dales of Yorkshire, custom prescribed the proper gifts to bestow upon such an occasion ; an egg signifying the necessities of life, a little salt (the luxuries), a silver coin (money or riches), and a match (to light the way to heaven), but only very old fashioned folk in very out-of-the-way places follow such ancient observances.

Marriage customs are many and various although seldom very remarkable. Holding up the bridal party on its way from church is still sometimes practised, a flower decked rope barring the road until the bridegroom pays toll. The bestowal of the Dunmow Flitch is a reminder of old methods of rewarding wedded bliss, while "rough music" gives a hint of more drastic ways of rebuking faithless and quarrelsome couples. Superstitious rites connected with death linger in the by-ways ; some will touch the corpse so that it shall not haunt their dreams, and probably the majority of bee-keepers tell the hives when a death occurs, sometimes draping them as well. Only very occasionally is a farmer borne to his last rest on his best waggon, and Abbotts Ann, near Andover, is apparently the only place where a virgin's crown is carried before the coffin of a young girl.

The work of the husbandman was permeated with customs, some almost as old as his calling, but modern methods, the loosening of the ties between master and man, and the restriction of religion to the Sabbath, to say nothing of more than four years of war, are rapidly bringing them to an end. Most of the feastings and rejoicings of the agricultural year have lapsed or are shadowy reflections of their former importance, observed by a few of the older generation or revived artificially by a lover of the past. The shilling clinching the bargain and ratifying the engagement is still regarded as binding as an agreement by a few greybeards, but is unlikely to survive the minimum wage. Formerly this was known as God's Penny, especially in the north ; on the Sunday after Hiring Fair, the labourers of Teesdale and Weardale usually attended church to give the coin to the collection. In Essex, the custom of dipping those taking part in the sheep-washing for the first time does not seem to be quite extinct, and in some localities sheep are still counted by the numerals of our British ancestors.

Custom often prescribes the date on which certain work must be done ; old country almanacks are full of them, and to this day the Devon allotment holder regards Good Friday as the only proper day to start planting " tetties," whereas in the north they usually say no more unlucky day for the use of spade or hammer could be imagined. Plough Monday merrymakings have not quite died out, but the day is no longer of importance to Church or agriculture. A special sermon is preached at Melton Mowbray, and Yorkshire farmers with their teams and ploughs sometimes visit a new-comer to help him on with his cultivation. At Whitby, the young men come in to celebrate the Plough Stots as of old, and the Plough Bullockers occasionally drive their decorated plough through the villages of Derbyshire, to the detriment of those who refuse largesse, but the plough has gone from the base of the church tower.

Haysel is now practically the first of rural festivals although no longer specially connected with St. Barnabas Day, but when the maiden arrives to help with the raking she hopes her lover will not forget to make " sweet hay," provided this by-product of the meadows is known in her neighbourhood. The frolickings of haytime were as nothing compared with those of corn harvest, but the old observances have disappeared one by one. The gleaning bell still sounds across the stubble fields of several parishes in Essex and other counties, but it is many years since it served any purpose, and Whalton, Northumberland, must be the last place where the Kern Doll is fashioned every year. Even there it serves only as a decoration for the church, yet many recollect when " Crying the neck " or " mare " resounded over the fields as harvest drew to a close and the harvest supper in kitchen or barn was the regular thing.

Customs in which the Church took part have largely lapsed and when saints' days and religious festivals are not forgotten the celebrations are mostly secular and often degraded to base purposes. Whitsun Ales, jollifications under supervision to prevent undue license, which helped to raise parish funds, disappeared long ago, but the Revel or Feast in connection with the dedication of the village church is remembered in the west and north. Rush-bearing reminds us of the time when the church was regularly strewn with rushes, Ambleside providing the best known example, though Shenington in Oxfordshire keeps up the custom, and at Old Weston, Hunts, the church is annually strewn with hay from a field long ago left for the purpose. Dressing the Wells is also connected with the Church, but most of these observances are revivals, that at Tissington being the most famous

Hardly a village is without its benefaction, though in order to meet modern needs it is often applied in a manner that might not earn the approval of its founder. Sums of money or bread or even articles of clothing are distributed at the church door or over the tomb of the benefactor at Christmas, Easter, or other festival, on the day of the saint to whom the church is dedicated, or on the birth or death anniversary of the donor. Such are the Lenten Doles of bread and herrings at Dronfield, the Tichborne dole, the Biddenden Cakes, Throwing for Maid's Money at Guildford, and Forty Shilling Day at Wotton, near Dorking. Land left for the purpose often provides the necessary income, and the tenant may be determined in the old manner, as by candle auction at Tatsworth, Chard, and Upwey, or by drawing balls at Yarnton.

If the wayfarer be learned in ancient charities he will know where and when to present himself for a free snack or meal, as that at Winchester is not the only one. The country considers eating and drinking of some little importance, and districts are still famous for special dishes and feasts, but they tend to disappear. Paignton has given up its wonderful pudding, said to take seven years making, seven baking, and seven eating, and Denby Dale seems to have allowed Peace Day to pass without serving one of its mammoth pies. Everyone has heard of Colchester's Oyster Feast, and cherry pie feasts have not quite vanished from rural Buckingham, while the Devil gives Cornwall a wide berth lest he be put into a pasty. Geese still diminish in number at Michaelmas, but whether the Armada has anything to do with it is another matter ; a stubble goose was regarded as a delicacy, and as tenants were accustomed to propitiate their lords with a gift at quarter day what more appropriate than a goose at Michaelmas ?

Mention of landlord and tenant is a reminder that the audit dinner is one of the few remaining customs showing the dependence of one on the other, and the collection of Wroth Money on Knightlow Hill is a survival of Feudalism. The Hocktide celebrations at Hungerford also consist largely of Feudal observances though Courts Baron and Leet are still convened in many places. Up to the time of the modern urban and parish council many appointed the officials, who were by no means merely burlesque sinecures such as pig drivers and ale tasters, once offices of importance. A few still regulate a market or possess some other right, and even a County Council has had its jurisdiction challenged by a Court jealous of its ancient privileges. The Court of Verderers has settled the affairs of the New Forest for centuries, and the Miners Court of Derby register decisions, its business being to decide the ownership of mines. The Barmaster of the High Peak convenes a court, and a grand jury is appointed to hear the evidence, a piece of fluor-spar being handed to the litigant who maintains his case.

Beating the bounds is another custom that appertains to local government and is carried out here and there with some of the old rites. Boys are usually the beaters, being provided with canes, which were formerly applied to them. Sometimes the mayor is bumped, and at Newbiggin, Northumberland, two new freeholders are bumped on the " dunting stone " on the moor. At Helston a clod of earth and a sprig of hawthorn are placed on the boundary stones and beaten by boys at a given signal.

Curfew, comparatively common as it is, serves more for the purpose of setting watches than sending people scurrying home to douse their fires ; at Ripon it is accompanied by the blowing of the Wakeman's Horn, for the same purpose, but is of older date. Another horn blowing custom exists at Bainbridge where it is sounded at nine o'clock to guide anyone lost on the surrounding fells, much as church bells are rung elsewhere to guide people home from market as at Kirton-in-Lindsey.

In old rural calendars the feasts with all the customs they include jostle one another on every page ; most are hardly remembered, but a description of those which remain would fill a good sized volume. The long ugly valentine may be seen lurking in untidy shops in little towns or hiding among the commodities of the village store. In Norfolk, the day is sometimes honoured by giving oranges or more substantial gifts to children, the donor placing them on the threshold and departing hastily after knocking. Shrovetide football flourishes at Alnwick, Sedgefield, and Chester-le-Street among other places, Cornwall preferring a hurling match. Often started with ancient ceremony, these rough and tumble games are not suited to the weak and delicate. The " pancake bell " may still be heard, and many a Midland housewife throws the first pancake to the hens to ensure good luck and plenty of eggs.

Mothering Sunday is only a name, but its special wafers were made at Chilbolton a few years ago, and frumenty is eaten in Gloucestershire, while Simnel cakes may be obtained in places. Carlin Sunday has a few devotees, and Palm Sunday is associated with the eating of figs in

197

THE HOBBY-HORSE WITH ITS ATTENDANTS—MAY DAY AT PADSTOW, CORNWALL.

NEW FOREST CHRISTMAS MUMMERS. *Left to right*—Johnny Jack, the Doctor, King of Egypt, Valiant Soldier, Turkey Snipe, St. George, Father Christmas.

" Room, room, brave gallants, there shall be shown
The most dreadful battle that ever was known."

THE HEADINGTON TROOP OF MORRIS DANCERS.

MORRIS DANCERS.

"With bells on legs, and napkins clean unto your shoulders tide,
With scarfs and garters as you please, and hey for our town cry'd."

CHOOSING TENANTS FOR MEADOW LAND AT YARNTON, OXFORDSHIRE.

MAY DAY CELEBRATIONS—COWLEY ST. JOHN, OXFORD.

BEATING THE BOUNDS—ST. MICHAEL'S, OXFORD.

Bucks and Northants, and maybe people still climb Silbury Hill to consume fig cakes washed down by draughts drawn from a spring below. The young men of West Somerset were wont to ascend Dunkery Beacon on Easter morn to obtain good luck in love and work, and old Devon people can tell of simple folk who visited the nearest height to watch the sun dance. Easter " Peace Egging," on the lines of Christmas Mumming, is not quite extinct in West Yorkshire, nor " heaving " altogether forgotten in Lancashire. Easter Monday at Hallaton sees the annual scramble for the two hare pies and two dozen penny loaves, followed by a rough game of football with a small beer barrel.

May is probably richer in seasonal customs than any month, but even they are dying and some are revivals. However, there are may-poles which have survived Puritan fanaticism and modern disdain, but most of the May Queens are children who cannot show a long unbroken line of ancestors, the best known being elected at Knutsford and Flore. Morris dancing, usually by children, is a feature of the more elaborate May-day pageant, and in the Midlands, Morris troupes proper perform on special occasions, giving the old dances whose figures are said to show traces of their pagan origin. Morris dancing is supposed to have had a Moorish origin and to have been introduced into England from Spain by John of Gaunt.

Children sometimes carry round the May Garland, and at Padstow and Minehead the hobby-horse still prances through the streets on May the first. The Cornish town possesses the most fearsome beast, but Minehead prolongs the revels and its " horse " sprouts a lengthy tail of knotted rope to chide those who overlook the money box, of old a boot adding emphasis to the appeal. A hobby-horse is also one of the characters in the famous Horn Dance at Abbots Bromley later in the year, its unusually elaborate pantomime being supposed to represent the forest charter of Henry III. Helston's Furry Dance on Old May-day is known by all, and Castleton, perhaps wisely considering our climate, is also late in its celebrations, for it is on Oak Apple Day that with fitting ceremony its garland is borne to the church and hoisted to the top of the tower. On Garland Day (May 13th) at Abbotsbury, Dorset, flowers are taken out to sea as an offering to Neptune, and some of the fishers on the same Chesil Beach bury cakes and ale to ensure good luck in their calling, as true a survival of pagan times as can be found.

Kingsteignton, Devon, annually roasts its lamb on Whit-Tuesday, but no longer in the bed of the dammed stream as ancient story and custom dictate.   Watching the sun rise over Stonehenge on Midsummer-day appears to be a modern practice, but the lighting of Whalton's Baal Fire on its eve probably goes back to pre-historic days.   The wood is brought by hand, and the children dance round the completed pyre while their elders foot it on the green to music provided by a fiddler. Formerly the need fire was lit in north-western dales when cattle disease worked havoc with the herds, which were driven through the smoke as an antidote, the kindling brand being passed from farm to farm in a westerly direction as quickly as possible, the wood being piled ready.

Children in Shropshire, Cheshire, and Stafford still go round " souling " in November, but apparently the cakes are made no longer, and the war has finally killed the great doings at Lewes, Bridgwater, and elsewhere on Guy Fawkes night.   One of the most curious customs of unexplained origin is that known as Salmon Sunday observed at Paythorne Bridge, near Gisburn, on the Sunday nearest November 20th.   People from Ribblesdale and beyond, repair to the bridge to gaze into the river to watch the salmon, provided they honour the day, not invariably the case.

The Christmas festivities overshadow other December customs and survivals.   Soon after the clock has ushered in the last month of the year Colchester announces the arrival of winter by proclamation of the crier, and on St. Nicholas' Day, Berden, at the other side of Essex, elects a boy bishop as the choirs of cathedrals and large churches used to do. Most of the old rites of Christmastide are lapsing and its folklore is being forgotten.   Comparatively few villages now possess a regular carol party ;   the death of the old fiddler broke up some and the war finally dispersed many others.   It is much to be regretted, for many an ancient carol was confined to one district and all such will be speedily forgotten, while the passing of these customs is destroying the old atmosphere of the season.   Where will you find the yule log and what West Country farmer dreams of wassailing his apple trees ?   The Mummers with their rhyming play of St. George and the Turkish Knight, often with famous historical characters interpolated or substituted for others, have taken the stage for the last time in most parts, but here and there a travesty is given by children.   It lingers in the New Forest where a troupe of boys act the old folk play with wonderful vigour, and a band of young men used to perform it rather shamefacedly.

Then there are the customs which are really superstitions—cures and the like—such as passing children through a split ash for hernia, creeping through a holed stone to cure toothache, repeating a verse from

Ezekiel to stop bleeding, and charms extensively used although casual acquaintance seldom reveals them. Belief in the evil eye lingers to a surprising extent, and white witches can usually be consulted by the faithful—how could the seventh son of a seventh son fail to exercise his powers? Some of the ancient rites of black magic are practised in secret, and the harridan of evil reputation must surely be a witch!

Omens and portents exist almost everywhere if one has the knowledge to read them, and weather lore is a mixture of unconscious observation and curious belief. Birds and plants bring good or ill luck according to circumstances, and almost every ordinary activity of life is best carried on or avoided at certain times and seasons. May, especially, is a month in which one should act circumspectly. How can good fortune be looked for if clothes are washed on May-day, or brooms are bought before the month is out? Worse still, if white-washing be done during May, death will soon visit the house!

Fairies may have gone out of fashion, and ghosts and goblins may not be so fearful or numerous, but nearly everywhere is a secluded dell or shady cross-way where the ghost of murderer or victim stalks at certain hours. Who knows what terrifying apparition may be encountered in lonely lanes after dusk has fallen? Perchance the padfoot, that horrid creature like a black pig with the speed of a hound, which causes the death of all it overtakes. There are heaths and hills and marshes where the Black Huntsman with his fiery-eyed pack hunts on stormy nights, and estates to which some previous owner returns at midnight to follow again his beloved chase. Black dogs and phantom coaches pass silently along deserted highways or come up the drive of the ancient mansion to summon an inmate to the next world, and white rabbits which do not belong to earth cause the death of those who would shoot them.

What neighbourhood is without its story of buried treasure; is there a wild heath which does not possess a pile of ruins or deserted house remembered as an inn of evil repute where travellers were murdered, and who has not heard of the abandoned hamlet depopulated by a visitation of the plague introduced by a bale of old clothes? Ancient wayside springs must possess some virtue, caverns surely always communicate with places far away, old houses and ruined castles cannot exist without secret passages and dungeons about which many wonderful things are spoken, and there are spots where story says grass will never grow.

Countryfolk can explain many things in a manner which shocks the scientist, the origins of the names of plants, the habits of birds and beasts, weather phenomena, and so on, but the present generation knows practically nothing of the virtues of plants and herbs. Education seems to have been more successful in eradicating useful knowledge than in destroying superstition. Folklore is too vast and intricate a subject to be lightly dismissed, but to ignore it altogether is to omit mention of an influence which affects rural life to a considerable extent. Beliefs and practices are often more than mere superstition, being relics of the days when faith was deep and remedies simple, and a sympathetic study of their origin and reason helps us to understand the contradictory and baffling personality of the countryman.

XIV

RELIGIOUS LIFE—PAST AND PRESENT.

ONE of the most characteristic features of the countryside is the spire of the village church peeping above its surrounding trees, a symbol of the faith of the land and its ordered life. In some respects still the centre of village life, the Church is no longer the sum of its existence, watching over the people from the cradle to the grave, ministering to them in days of sickness and poverty, presiding over their festivities, settling their disputes. The old order has not quite passed away ; great as have been the changes and numerous the separations, the Church carries on many of her ancient duties, and the influence she once possessed is still to be traced in some methods of local government. Here and there, the remains of monastery or priory—some preserved and adapted as private mansions, others mere fragmentary ruins—serve as monuments to the day when all thought alike in religious matters.

Country churches range from the fine edifice rebuilt or enlarged by wealthy men as thank-offerings for success, to the tiny building sufficient for the needs of a small community among the hills. Many stand within a park, showing the old connection between church and manor, and the ruinous chapel, sometimes used as a shed among the out-buildings of an old farm, may date from the religious revival of late mediæval times when hundreds of these private worshipping places were licensed. Most characteristic of all are the simple village churches, stamped with marks of the age in which they were built and often containing beautiful examples of long dead craftsmen's art. Simple they may be, but few can enter them without a feeling of reverence ; on Sunday the sounds of the country play a subdued accompaniment to the parson's voice, and the scents of the flowers and hayfield wafted through the open door are as sweet as the finest incense.

Every church, and many a chapel, with their furnishing has something to say about the religious life of rural England. We can trace periods of great religious activity when churches were rebuilt and beautified, and, less easily, the times of neglect when little attention was paid to the Faith. Their embellishments and tombs reveal something of their builders and the people who worshipped within their walls.   Memorials and carvings and wall paintings reflect the art and taste of other days, and in the manner in which these were treated by subsequent generations we can glean a little of the changes in religious ideals.   Defaced statues and empty niches show the reforming zeal and austerity of the Puritans, mutilated screens and bench ends hidden in odd corners testify to unfortunately conceived " restorations."   Sometimes we come across evidence that the old ways are being reborn, for village carvers are seeking to replace by their own handiwork what vandalism has destroyed.

The furnishings throw some light on the religious customs and manners of our forefathers, but the church interior familiar to the Georgian era is rapidly disappearing.   The high pews are going, and the family pew, that occasionally sumptuous enclosure where the elders could snooze and the children play unseen, is almost a thing of the past. Some search must be made for the three-decker pulpit ; the hour glass, always coupled with long sermons, is a very rare fitting, even its bracket having mostly disappeared ; and it is still more difficult to find a church that cherishes its sluggard's wand.   However, the tongs with which intruding curs and wandering sheep-dogs were removed, repose in a few vestries, and perhaps a pitch-pipe or a mouldering mechanical organ is preserved.   The gallery where the choir sat has not invariably been improved away, and the list of charities on the wall shows how the poor were relieved.   Maybe a century or more old notice recalls an era already past when local government centred round the Church and the building itself was used for many secular purposes.

Would we learn to understand the close connection between Church and people, we have only to study the Registers and Churchwardens Accounts.   The former are not always bald entries of baptisms, marriages, and burials, but frequently contain notes on all sorts of matters, perhaps the astrological prospects of the infant christened or remarks on the character of the dead.   They recall the changes of the Commonwealth, and tell of times when burial in woollen garments was enforced under a penalty of £5, and when each entry paid a tax according to the value of property held.   It is also interesting to find that neither high-sounding names nor the custom of calling children after famous people is a modern fad.

Churchwardens' Accounts though primarily a list of parish expenses make us understand how inextricably mixed were religious and secular *affairs.   The matters which came under the ken of the churchwardens* concern every side of local government, even to payment for the destruction of noxious fowls and vermin—including, let it be whispered lest sportsmen overhear, the fox—and entries prove that problems supposed to be confined to our own times exercised the minds of our ancestors.

Besides particulars relating to Church matters such as rehanging or recasting the bells, we find the number and equipment of soldiers provided by the parish, the names of pauper apprentices, and the cost of sending people to London to be touched for the " king's evil." Fines for non-attendance at church figure among the receipts, as do sums in default of penance, which was imposed until comparatively recent times. Now and again a note states that a charity cannot be disbursed as the money has been lent to a needy person or advanced for a special purpose and not repaid to time. The cost of drink occurs more often than modern opinion considers seemly, wine for a visiting bishop or rural dean, wine at a wedding, and drink for workmen, in fact it would appear that wages were sometimes paid in ale.

Churchwardens did not have it their own way entirely as the rural dean made periodical visits to make sure they were attending to their duties. Fines were inflicted for remissness, though they do not appear to have come out of the churchwardens' own pockets or we should not expect to read :—" For repairing Church gates and hedges and Byble iijs. vid." The lady churchwarden is not an innovation as she will be found some four centuries ago, apparently carrying out her duties by deputy. Although shorn of his former public duties the modern churchwarden is usually a man of substance and influence in the parish, often one who has served for many years, perhaps following in the footsteps of his father.

The place of the parson in local government is also less important, some of his public offices being bestowed more from courtesy than legal right, but if he had nothing more to do than walk circumspectly, speak tactfully, and endeavour to be all things to all men his time would be fully occupied. Nominally head of the parish and its dominant personality, the poor man feels as clay in the hands of the potter when he stands before a masterly churchwarden, is told some home truths about the selection of hymns by a stubborn choir-master, or is lectured by a managing lady parishioner.

Much has been written about the country parson but it would require a whole library to recount what the village says about him. He is the perennial subject of criticism, deserved and undeserved, and is lucky indeed to be canonised before his departure, but then his successor will think of him with feelings far from saintly. He is expected to be the almoner of the needy, real and imaginary ; the purveyor of entertainment, suitable and popular ; and the provider of hospitality. He is constantly requested to deliver judgments that on no account must give offence, to find servants or holiday appartments, to supply guidebook information, trace " ancestors " who never saw his parish, and answer abstruse antiquarian or geological questions concerning his district.

His virtues and his failings are as old as the Church itself, and the abuses connected with him are by no means new. The pluralist rector is not beloved, and the suggested amalgamation of rural parishes seems likely to be regarded in the same light. Long ago monasteries played the pluralist parson, adding to their incomes by pocketing the difference between the tithes they received and the stipend paid the officiating priest. The worthy deeds of saintly parsons are known only to their parish and generation, but the scandalous acts of a single evil-doer render his village notorious evermore. Criticism of the sporting parson is usually more rife outside than inside his parish, especially if he hunt, for those who know him not, picture him as a Froude, whereas, unless he be remiss in his proper duties he has probably greater influence than the vicar who finds his recreations in his study. Often he is the mainstay of the cricket eleven, and you may be sure the more manly he is the less he has to complain of a one-sided congregation.

His parsonage may be a fine old building, originally a manor house perhaps, but so expensive to maintain he lets it when possible and lives in a more modest dwelling. Some with dormer windows peeping from eyelids of thatch seem fitting homes for those whose work lies among an agricultural population, and some live in history on account of their connection with famous men. But they hide many tragedies, tragedies of poverty and misunderstanding, for too often clever men, earnest men, even tactful men, fail to get on with their parishioners, and the Enabling Act is not going to smooth the path of the village parson.

The country clergyman is born not made ; if he fail to make allowance for prejudice or to understand the countryman's turn of mind he labours in vain, and the breach between him and his flock becomes wider the longer he remains. However, in spite of his many troubles he lives to a hale old age ; rectorships of thirty years are too numerous to be uncommon, and forty years are not rare, but the record must be held by the late Rev. W. W. Wingfield, vicar of Gulval for close on seventy-four years, who conducted services and preached sermons at ninety-four.

PROCESSION TO THE FIELDS, ABINGDON.

A VILLAGE FUNERAL.

A LITTLE CHURCH AMID THE FELLS—KENTMERE.

"To what base uses."
KERSEY CHAPEL, SUFFOLK.

THE MISSIONARY PLAY.

A CORPUS CHRISTI PROCESSION, COWLEY, OXFORD.

212

THE SQUIRES MONUMENT—PILTON CHURCH, NORTH DEVON.

THE SEXTON.

Both parson and his flock must have felt pleased when the tithes were commuted, for the old method of collecting them in kind was vexatious and occasionally a matter of warfare. At harvest, farmers had to give notice of reaping so that the proctor might mark each tenth stook, and it was not easy to change the day. Every tenth animal and vegetable were set aside, and the milk of every tenth day fetched. The easygoing rector often received less than his due, and the rapacious one managed to obtain more than he should. Now and again old wills mention sums of money left to clergymen in lieu of unpaid tithes, " conscience money " of the period.

The parson's wife has every opportunity of sharing in his trials and difficulties or rejoicing in his rarer triumphs. She must suit her conduct to the idiosyncrasies of the parish, and if she would help him in his work must be neither tactless nor what the village terms worldly, but if she understand human nature and be sympathetic her influence is greater than it appears. She, too, must be born to the work, though experience widens outlook and increases discernment, but even then she may fail to make her husband a success ; if she be lacking in essential qualifications she is a sad handicap to the best of parsons, but being like him judged by a variety of opinions can never hope to please all.

The curate is confined to large parishes with dependent churches and to those where the vicar is too infirm to carry out his duties. Usually he has to face even more censorious criticism, and his landlady is apt to tell tales out of school. Sometimes those who condemn him add an excuse, " But then, he is only learning his trade," and sometimes he endears himself so much to all that they ask for his preferment when the living falls vacant.

A wealth of anecdote that will not be quickly dissipated is practically all that now represents the parish clerk. His place in modern story has been taken by the verger, or sometimes the beadle, more often than not a labourer who doffs his cap to the congregation on week-days and plays the autocrat on Sunday. He, also, is a character, and often when old and past work haunts his beloved church ; he is never so happy as when showing round the stranger, reciting its history—or what he regards as its history—pointing out everything to be seen and repeating ancient story, but a handsome tip does not allay his contempt for those who refuse to hear the recital to its appointed end.

However, this race of verger, resplendent or sombre of garb and bearing a wand of office, is dying out ; year by year, a new tombstone or a fresh inscription on an old one registers a decrease in their numbers.

His duties were frequently combined with those of the sexton, who is also an " old ancient " of bowed white head, familiar with death and regarding grave digging as a hobby concerning which he is ever ready to discourse.   Verger and sexton have often been provided by the same family for generations, the office descending from father to son almost as a matter of right, and great is the general regret when the succession fails, but in these days of public cemeteries the sexton will soon have joined the parish clerk in the Shades.

Old folk remember the time when the musicians in the gallery led the singing.   Perchance, some thought more of their individual performance than the general effect, but the same can be said of the modern choir.   At least they helped to keep the folk in touch with the Church and handed on the traditional hymn tunes and carols, besides preserving old ballads and country dances.   Now a harmonium played by vicar's daughter or schoolmaster provides the music in tiny churches, but most villages aspire to a proper organ and surpliced choir.   At Muchelney the old clockwork instrument remained in use until 1907, and the well-worn story of an occasion when it refused to stop was duly related at the dedication of its modern successor.

On the Sabbath, peal still answers peal across the fields, and it is difficult to imagine a day of rejoicing without crash of melody from the church tower.   The bells welcome the bishop on his visitation or an heir to the squire, every self-respecting country girl demands a wedding peal, and the solemn passing bell announces that the community has lost another member whose identity can perhaps be guessed if the toll is followed by the " tellers."   England used to be known as the Ringing Isle, but in too many places old ringers lament that in spite of more leisure young recruits seldom appear in the belfry.   In the ringing chamber hang the rules, often in doggerel verse and sometimes very old, giving the penalties for late arrival, wrong notes, idle talking, bad language, and unseemly conduct, the jug of beer fine being much in evidence.

Visitors to rural parts declare that chimes are too insistent and disturb slumber, but they do not worry the villager, remind field workers of the passing time, and may recall old story or tradition.   Single bells rarely attract attention, and though numerous their origin is usually forgotten as they are often known by some modern name as Pancake Bell or Oven Bell.   Some of the daily bells are survivals of the Curfew or the morning and evening Ave bell ordered by Archbishop Arundel in 1399, which is still sometimes called the Gabriel Bell, as at Dunster in

Somerset, but often it is sounded in the morning only and serves as a call to work. Here and there an evening bell is rung during the winter months, originally a guidance to travellers when roads were few and difficult to traverse after dark. At times bells were rung for quite vulgar purposes as in a certain Devon parish where not so long ago a bell announced the local butcher had " pig's in'ards " for sale, or when a hasty jangle was a call to destroy a prowling fox.

Many regret that the Church no longer takes her old corporate part in the life of the community because most of the services connected with everyday activities have vanished or survive without their original significance ; but there are occasions when only those with no religion at all fail to attend, especially at times of public rejoicing and sorrowing when the Church if only for an hour or so becomes truly national once more. Yet, allowing for change of habits and thought, the religious life of the village shows more continuity than might be supposed, Church and Chapel continuing old services and customs, each in its own manner. The Revel or Wake or Feast is annually celebrated in some parts, especially Cornwall, where all denominations observe the festival. Harvest Thanksgiving services are arranged on different days so all may attend, as the Parish church is always full to overflowing.

The Church Ale is now regarded as indecorous, being replaced by the Church Social, but at least one rector has instituted a smoking concert from which ale in moderation is not banished, while some preside at dances and other amusements in the local hall. Feasting, of a very modest order, figures largely in the Church calendar. At the Choir Outing, most popular when the members choose their destination, the luncheon basket is of great importance. The Old Folks' Tea is a simple feast, but no one who has taken part in the Sunday School Treat, that anniversary which breeds a latitudinarianism among the young positively amazing, can deny the feasting that takes place. The Mothers' Outing is purely a pleasure trip, and the spirit of emulation rather overshadows the devotional side of those musical festivals in which the diocesan choirs take part. The miracle play is represented by entertainments organised to raise funds for Church purposes—sacred tableaux depicting episodes in the lives of prophets or saints, playlets and pageants founded on our religious history or successful missionary effort.

Rush bearing and well dressing remain as customs, but the plough light has gone from before the altar, though in some parishes Rogation-tide brings its procession into the fields. Selsey holds a service to ask a blessing on the fishing, and Gunwalloe one on All Souls' Day for those lost at sea. Painswick, among very few places, celebrates the annual Clipping the Church, when the congregation, clasping hands, forms a ring round the church while special hymns are sung, followed by a sermon from the chancel door. Some say " clipping " means embracing from the ring round the building, but others consider it the old word for naming as the service takes place on dedication day.

Few country people consider a marriage is a marriage unless it takes place in church though they may seldom attend at other times, but weddings are not accompanied by those riotous scenes once common in some districts. A village funeral remains pretty much what it was, a ceremony impressive for its simplicity, except when pretentiousness turns it into a show as is perhaps more frequently the case than formerly. The coffin is borne by comrades and followed by relatives and friends engaged in the same occupation. Occasionally a squire or farmer may be carried to his last resting-place in one of his own waggons, and here and there girls in white accompany the coffin of a young friend. The funeral tea, followed by an undignified scramble for the deceased's belongings, is still rather common in the cottages, and usually the bereaved relatives, attired in black from head to foot, attend church on the succeeding Sunday.

The chapel is as familiar as the parish church though it may be neither so ancient nor so ornamental, that at Horningsham, Wilts, being one of the oldest, but it may impress by its simplicity and associations. Who that has stood in the sequestered graveyard beside the Meeting House in the leafy bottom at Jordans has not understood something of the simple yet sturdy piety of those sorely persecuted Friends who came from near and far to worship in that plain room ?

Perhaps the humble preacher who gives up his Sunday rest to lead the service in distant village or hamlet is not quite so much in evidence though by no means extinct ; his homely eloquence has more effect on his congregation than that of a man better endowed, and his disappearance would be regretted. The Nonconformist minister has his definite place and duties and has done much to influence the parent Church for good. Some think there is always enmity between Church and Chapel, and true though it be in some unhappy places it is far from common and many differences have been bridged within recent years.

Cornwall is noted for an austere Wesleyanism with strictly Sabbatarian notions, and is also famous for its great anniversary services, of which the best known is that at Gwennap Pit. Wales also regards Sunday from the Puritan point of view, but opinion has become less

HORNINGSHAM CONGREGATIONAL CHAPEL, WILTS.
One of the oldest Nonconformist Churches in England.

THE VILLAGE CHURCH, WEST STAFFORD, DORSET.

A SURVIVAL FROM THE THREE-DECKER PERIOD—WHITBY
PARISH CHURCH.

TEA AT THE MOTHERS' OUTING—ARDLEIGH, ESSEX.

BRAY, BERKSHIRE.   WORTH, SUSSEX.

TO THE MEMORY OF THE DEAD.

rigid within recent years.   In remote places still untouched by broader views and modern indifference the old strict opinions still hold sway, and everwhere the sturdy but bent old man tramping several miles to chapel is a familiar figure.   Many districts are becoming quite accustomed to the sight of the monk and nun, and as Roman Catholic settlements with their monasteries and convents are too common to excite remark, their festivals attract many, and even the Corpus Christi procession seldom arouses controversy.   The Gospel Van sometimes draws up on the green, and the Salvation Army band visits the village, while the fanatic with a paint pot walks abroad at mysterious hours to scrawl his texts, preferably of a denunciatory order, on field gate and fence

Modern tendencies are influencing village as well as town, and the reasons that make some go from church to chapel or *vice versà* may make the cynic laugh ; where the parson is unpopular, even for insufficient reasons, religion may seem to be losing its hold on the people, but on the whole the Sunday of tradition survives in rural England.   Nothing is more characteristic of the country Sabbath than groups clad in their best proceeding by footpath and lane to the church whose bells ring out the call to service.   From wayside chapel comes the sound of hearty singing, and sometimes the passing wayfarer can join in a simple open air service at a distant cross-way.

XV

WAR TIME AND ITS AFTERMATH.

ONLY now that the War is receding into the past are we beginning to understand the upheaval it made in English life. The countryside was the last to feel its full effects and is likely to be the last to recover, for the results that followed in its train can only be ascertained after the experiences of many seasons.

The war brought change greater and more far-reaching than any wrought by decades of peace, and yet with it all things seem likely to settle down eventually very much as they were. The country is conservative with the conservatism born of experience, experience ingrained in the bone and blood of many generations. It is not averse to progress, but, having continually to face disappointment, it accepts only that which can be proved. During the bad times of the last quarter of the nineteenth century it had evolved, by toil and experiment, a system that provided a living, and was leading to quiet prosperity such as it had not known for many years, so that its initial slowness in adopting methods urged chiefly by those with little practical knowledge was caused neither by the stubbornness of ignorance nor opposition to progress.

Town and country have now so little in common it is necessary to labour this point, if the effort of rural England during the war is to receive its due. The countryman was asked to go counter to his better knowledge gained in years of experience, mortgage his future, and yield his cherished independence in everything he undertook. In this he was not singular, but the cultivator is, perforce, an individualist, because his living depends upon his powers of judgment and his knowledge of his own special circumstances, and he recognised he was engaging in a vast experiment.

It is difficult to give an adequate impression of this effort, and it is doubtful if the story will ever be told in full, being chiefly a matter of petty details of everyday work carried on at high pressure, against time, with inadequate and only partially skilled labour, and, like all country work, it differed largely in every district. There was nothing spectacular about it, and no special correspondents to show how all the tasks were correlated, or to turn commonplace doings into an epic. Those mostly concerned knew only what was being done in their own immediate neighbourhood, and few had sufficient imagination to comprehend the full import of their labours.

To attempt to set down the tale of progressive endeavour in chronological order would be futile, because only towards the end was there any semblance of a clear-cut plan. One rural industry after another was recalled to life or started into full activity after having almost ceased to exist. Half-forgotten crops became familiar again ; land that had reverted to thicket was cleared and cultivated ; the ancient virtues of wild flowers and fruits were re-discovered or new uses found for them ; in fact, most of the country arts and crafts of bygone eras were reborn.

It was not entirely a story of rural activities, for the highly organised factory sprang up on waste land and in village, which grew into a town almost in a night ; new lines of railway were constructed, and little used branches were torn up and transported to places where they would be of better use ; camps and training centres scattered their huts and parade grounds over ploughland and pasture of unmilitary districts which hitherto had hardly ever seen a battalion in full marching order. Owing to the many diversities of rural England, this gradual awakening into lusty life resembled the advance of the tide on a deeply-fretted shore, there with a slow but steady flow, elsewhere in little rushes with an ebb between. It came about slowly, almost imperceptibly, then with ever increasing speed and vigour, until at the Armistice all except the few who wilfully stood aside were engaged in some task to further the common cause.

If the full effects of the war were not immediately felt by country folk it was not a little due to the authorities, who were even slower in recognizing the vital importance of agriculture, which was almost the last of our great industries to be mobilized. Then, when the task was started it seemed well-nigh hopeless, for agriculture had been neglected for years and left to work out its own salvation, instead of being fostered and organized as a national necessity. The early days of the war increased disorganization, so that when the call came the countryside was depleted of skilled labour and horses, with means of transport greatly reduced, short of fertilizers and seed, lacking in sufficient machinery, especially in parts where climate and other considerations had reduced tillage to a minimum, and the land foul from want of attention, altogether unfitted in almost every way for the enormous extension of cultivation imperatively demanded.

Farmers had endless difficulties to overcome. Extra horses were obtained only by persistent correspondence and the tiresome process of filling up sheaves of forms ; after additional land had been prepared, favourable sowing weather was lost by the non-arrival of seed ; and on top of all control did not ease the situation. The farmer's carts were checked and numbered, his stacks bore the broad arrow and initials of the War Department, he could not cut a bale of his own hay without permission, and inspectors were ever at his heels or examining his books. The price of the produce he sold was fixed, but he had to pay market rates for his raw material, damaged corn for stock feeding being worth more than sound grain.

The independent and energetic preferred to do their best without appealing for Government aid, and all responded to the call with such readiness and determination that the acreage of corn and potatoes often exceeded the quota demanded. For example, Cornwall, never considered very suitable for wheat growing, raised its acreage from 19,000 in 1916 to 57,000 in 1918 ; in part of another county largely devoted to dairying mixed farmers were asked to put 75 per cent. of their arable into corn and potatoes, and graziers to break up 30 per cent. of their grass. As a result, 116,000 acres of corn were cultivated instead of the quota of 102,000, while a comparatively small district elsewhere expanded the demanded 429 acres of potatoes into 2,204 acres.

Comparatively few have realized the extent of this great effort, and it is to be regretted that fewer still have realized its true value and significance. It was not achieved without a display of that optimism and tenacity in face of discouragement which have so often helped the countryman to overcome a difficulty, and even then the effort might have failed without the co-operation of all, and the loyal aid of many from the towns who took up work to which they had hitherto been strangers. Masters worked harder and for longer hours than their labourers, managing, somehow, to instruct the neophytes as well ; daughters took the place of serving brothers; boys did the work of men; and veterans proved that skill reduces the handicap of age.

As time revealed the need, all answered the call to produce more food, until at last almost every piece of ground that could be coaxed into growing something bore its crop. Landowners ploughed up their parks ; the squire's garden became a potato patch ; the cottager added to his allotment when possible, wives and young sons often doing wonders in the absence of the head of the family, while the parson gave a helping hand. In the holidays Public Schoolboys cleared land that had reverted

LANDWORKERS CLEARING UNDERGROWTH, BETTWS-Y-COED.

WELCOME GUESTS AT ALL FETES.

PULLING FLAX—A HOLIDAY TASK FOR MANY CITY WORKERS.

to waste, and planted potatoes or prepared the ground for Land Girls who followed them, and the crop thus sowed was gathered by village women and boys from the local elementary schools.

The pretty blue flower of flax covered fields where it had been forgotten, its gathering being a holiday task for hundreds from the towns, mostly women and girls. Land Girls travelled about baling hay, which was then stacked in long lines beside convenient sidings at quiet wayside stations. Women also assisted in forestry work, which became a great industry in woodland districts, mainly carried on by Canadians and Portuguese, and later by German prisoners. Settlements of log huts, more familiar to the backwoods of America than the ordered landscapes of England, sprang up in clearings and on hillsides, and coverts once sacred to the pheasant were transformed into deserts of mud mottled with tree stumps and mountains of sawdust. In alcoves formed of stacks of timber, beside roads and tracks to facilitate loading, saws driven by oil engines speedily converted trees into pit props and trench supports. Charcoal burners found they were still part of war's auxiliary corps.

Each district, in fact, had its war industries, some common to all, others peculiar to special places and known only to few. Local millers struggling against competition from big firms at the ports were busier than they had been for years, grinding local corn, with additions against which their souls revolted; but this prosperity was bought at the heavy price of Control that turned the miller's hair grey before his years. The stone crushers roared all day at the edge of quarries, yielding good road metal; the " braiders " or net makers of Dorset had no lack of work, as nets were wanted for various purposes, and some disabled soldiers made shell-carrying cases of cane and osiers. In hilly moorland country the collection of sphagnum moss for dressings made an excuse for a walk, and children enjoyed gathering blackberries and whortleberries or the wild flowers required for dyes and drugs, a few enterprising individuals even growing them especially.

Some carried on their work familiar with the buzz of hostile aircraft and the sound of exploding bomb, while a few watched thrilling battles in the air and rejoiced to see the destruction of enemy raiders. But our countryfolk were not easily daunted; one dame of eighty, visiting relatives in London during an air raid, ran to the window to see, for " I mightn't get the chance again " she said when warned of danger. Villages astride highways to certain ports were too well acquainted with convoys of ambulances filled with wounded, and those dwelling coast-wise sometimes came into close touch with the horrors of war as waged by an unscrupulous foe, when stricken hospital and merchant ship staggered to the nearest haven or were beached.

Sports and amusements were almost forgotten. The cricket field folded the flock, and lawn tennis parties gave way to gatherings where sandbags and bandages were made. Hounds met but seldom, then mainly to keep down foxes or provide sport for officers home on leave. The local race meeting disappeared in company with the agricultural show, and even dances no longer enlivened winter evenings. But as convalescent wounded increased and toil grew more continuous some relaxation became necessary. The " Blue Boys " gave concerts in aid of their own hospitals, and they were honoured guests at country house parties, athletic meetings, and other entertainments arranged in aid of war funds, which provided the excuse for all amusements.

As market day became largely a shadow of its former self, townlet and village had to rely on their own resources to an extent they had almost forgotten, and struggling shopkeepers learned that war brought unexpected advantages, and that even Control had its profitable side. Childish voices no longer sounded around the secluded cot, for father had been called up and mother gone to parents with her children ; many country houses were empty or let to strangers, as their inmates were scattered far and wide—in the field, hospital, or canteen. Whispers from little-frequented beaches related that houses were provisioned and partly furnished with spoils washed ashore from victims of the lurking submarine, though perhaps these stories must be put beside the rumours of disaster spread by " travellers " or passing " motorists " whose identities were difficult to trace.

Country occupations are so rarely characterised by bustling activity and so seldom concentrated in one spot that the busiest season may appear a time of sleepy ease. Thus, when all were pre-occupied with their own work the common effort was revealed only by some chance impression. The details which meant so much are fast fading from memory, and war time in the country is becoming a matter of a few vivid recollections.

The dignified old squire pushing a farm wheelbarrow stood for the willing co-operation of all, as did the lady helping the cottage woman to wash dishes in the village hospital, while the reverse was illustrated by the retired shop-keeper who expressed her willingness to assist provided she were given no " menial work." A Belgian Army mobilisation notice in French and Flemish posted on walls in the edge of beyond, and the request of a flustered cottage woman that a passing stranger would accompany her down the road, as she was nervous of a loitering Portuguese woodman, showed how the world—and at least his Belgian wife—came to English byways. Knowing the treatment of our own men, the sight of German prisoners—" Frying-pan Jacks," as children sometimes called them on account of the red patches on their clothes— comfortably driving to work smoking cigarettes, was bound to remain in memory, together with the dark looks and muttered exclamations of the women who saw them pass daily.

Close observation was not required to notice the absence of men of recruitable age, and the spectacle of a farmer carrying a basket covered with a cloth—he was taking the dinner to an old labourer— emphasised how necessary it was to propitiate all who knew their work. And last war recollection of all—that dark evening when groups waited about the village street to learn whether the rumour of an armistice was true. The ringers stood at their doors ready to hasten to the belfry, and the lantern of the passing cowman flashed weird shadows of man and beast on cottage walls, a memory more persistent than that of a few days later when the official announcement arrived, long after it was known through those devious channels by which news and rumours still reach the quiet corners.

To the villagers it signified immediate peace and settling down to the old round, with all the advantages war had brought without its cares and inconveniences. So they hung out faded flags hastily retrieved from hiding places, and the women talked in excited groups and the boys made much noise, while after dark a few makeshift bonfires of hedge clippings flared fitfully for a few minutes in defiance of regulations they hardly infringed. At the Armistice service the old church was full to overflowing, for all joined in thanksgiving, the Nonconformist minister co-operating with orthodox parson by reading the Lessons.

At first the hesitating appearance of the promised wonders of the new world caused disappointment ; but war had proved the vital impor- tance of country crafts and brought prosperity to many, so when the strain relaxed all enjoyed the unaccustomed liberty and freely indulged in amusements, some villages losing count of the " Victory dances " held in their neighbourhood. Preparations for the fitting celebration of Peace Day and discussions about war memorials kept minor dis- contents within bounds. The continuance of Control was irksome— though even when tightest its humours and its follies and its incon- sistencies caused many a laugh—but most farmers admitted they had as much reason to be thankful as to grumble, and anticipated the day when its dead hand would be removed.

Peace Day celebrations came somwhat as an anti-climax, but miniature reviews, with distributions of medals, that took place in the market squares of some ancient townlets, and the time-honoured costume parades, sports meetings, and public teas, with fireworks and bonfires, were enjoyed with the usual gusto. They relieved to some extent the unpleasant awakening to the fact that the unravelling of tangled skeins and the mending of broken threads require as much patient toil and combined devotion as the waging of a great war.

This period of transition was utilised as an opportunity for a preliminary stock-taking, clearing up what might be termed the débris of war. The superficial changes could be quickly recognised, but the far-reaching results developed slowly, and it will be long before their actions and re-actions cease, and the pendulum of country life settles down to its old steady beat. Work continued on the lines to which all had grown accustomed, but new difficulties had to be faced almost at once, difficulties less easily surmounted, because, the common danger having passed, the general welfare, the bond that gives the community strength, was forgotten amidst clashing interests.

As the months went by countryfolk returned by twos and threes to dishevelled surroundings. Their cottages were sadly in need of repair owing to high prices and the lack of labour, especially where heavy lorry traffic had not improved leaning walls or chimneys nor made roofs of tiles and small slates more watertight. A neighbouring wood was represented by an ugly scar, and the worst field track could give points to the deeply-rutted and pitted road by which it was approached. Some, dwelling on the downlands, found the old open sheep runs divided into wired pastures, a result of lack of shepherds and high wages it was said.

Certainly, many modern innovations and conveniences had arrived in a night as it were. Trains might be few and crowded and the old carrier retired, but the motor bus, maybe a London style two-decker, now ran two journeys where he made one and served villages far from the railway. Even tiny places possessed a small motor van for market days. The whole of southern England appeared to have become a holiday district, for the motor char-a-banc was met everywhere, being a nuisance on narrow roads leading to beauty spots. All seemed bent on holiday, and many living within miles of a neighbourhood with the least claim on holiday-makers quickly accumulated small fortunes in catering for them.

LANDGIRLS BALING HAY.

BOYS WORKING IN THE HAYFIELD.

WOMEN IN THE FIELD, DEVONSHIRE.

MAKING BASKETS FOR CARRYING SHELLS.

THE WHEELWRIGHT'S WIFE GIVES A HAND AT TYRE COOLING.

FORESTRY GIRLS.

GERMAN PRISONERS SPRAYING POTATOES.

So the new England unfolded itself, and in spite of the drawbacks it seemed as if a golden age of happiness had indeed arrived, bringing high wages and short hours, with prosperity for all. But the older generation, who could see below the surface, shook their heads and regarded it as the froth on the stream in spate after a storm. Their native shrewdness and common sense recognised that it was too good to last, and that most of the activity consisted in overtaking arrears, that high prices nearly turned the scale against high wages, and that good money easily earned was having a demoralising, unsettling effect, and restlessness taking the place of steady endeavour. Worst of all, the war, followed by high taxation, had robbed the country of its natural leaders.

These forebodings were realised all too soon, for the schemes to ensure agricultural prosperity could not stand the cold blast of actuality, little attention having been paid to the advice of those who best understood country aspirations and difficulties. Thus did the old conditions begin to re-assert themselves. For many years rural problems have been considered from the townsman's point of view, which in the end will not benefit the city and ruin the countryside. Much has been heard of the garden city, which in reality means improving the town, not developing the country. In the last thirty years the agricultural area has decreased by two million acres, chiefly owing to urban extension, and garden cities would decrease it still more. In the same time the percentage of people living in rural districts has fallen by 7 per cent., from 28 per cent. to under 21 per cent.

Already much harm has been done by neglecting to consider what result legislation and taxation were likely to have on agricultural England. The hereditary landowning class, which enabled the country to weather many a storm, is being crushed out. Land and estate sales have multiplied during the past few years, bringing fresh difficulties and reducing the demand for labour. New faces are everywhere, on estate and in village, mainly people out of touch with country life who have served no apprenticeship to their duties and responsibilities. Farmers anxious to retain their holdings bought at the top of the boom and now find that mortgage holders are not so accommodating as the landowner, who would postpone taking his rent until better times and in some cases reduce it. Repairs and upkeep are expensive, and too often you may recognise the yeoman's farm by broken fences and rickety gates.

High costs have sadly interfered with the reconstruction so fondly expected, and proved the last straw that broke the back of many overburdened country gentlemen. When an estate owner has to pay over £124 to rethatch and repair four cottages bringing in less than £30 a year between them, and partial repairs to a small house cost over four and a half years' rent, how can the most generous landlord afford to do his duty to his tenants and dependants ? So every week estates are changing hands or being broken up, and if unpurchased are partly closed because their owners cannot maintain them properly.

A glance at history shows that time after time legislation to better conditions has defeated its object. If the agricultural problem be studied with sympathy and understanding, with a grasp of the interests of landowner, farmer, labourer, and all whose livelihood depends on them— an army of greater size and diversity than generally realised—a solution should be found. Manhood is more than wealth, and the general well-being of the nation superior to the theories of the doctrinaire or success of the politician. We are at the parting of the ways. Are we going to revive the country with its own industries and its own life, or are we going to spread the city over field and meadow ? Let us consider before we make a choice, let us be quite sure where we are going, for a false step cannot be retraced.

The country is more than a sporting ground for the wealthy, a place for a pleasant holiday, or a garden for the intensive cultivation of votes ; being constantly in touch with realities, it has a stabilising influence in these days of unrest and vague longings, it keeps alive the traditions of the nation and maintains the rugged strength which have made it great. Its ever-varying tasks preserve individuality, and are so interdependent that they form a bond that unites the interests of one to another, and show that each is necessary to the common weal.

The cities and great towns of England may be her brain, but the country is her soul. Allow the intellect to enslave the soul and you have cleverness without depth or humanity, and when the prolonged crisis arrives the lack of faith and unity brings inevitable collapse. Both acting in unison create the spirit which commands the world, whatever the circumstances.

INDEX

NOTE.—The use of black type figures denotes that the reference is to an illustration.

A

Abbots Bromley, **191**, 201
Abbotsbury, 201
Abbotts Ann, 193
Agriculture, 11, 12; number of workers, 2; implements, 11; Cicero on, 21; in bygone days, 40–42
Aldeburgh, 129
All Souls' Day, 202, 215
Almshouses, 80, **83**
Alnwick, 196
Amusements, 15, **142**, 144; 173 *et seq.*
Apparitions and Ghosts, 29, 203
Apprentices, Poor Law, 42, 207
Auctions, seasonal, 59, 116, 162; candle and curious, 195, **199**

B

Baal Fire, **192**, 202
Badminton, **26**
Bainbridge horn-blower, **189**, 196
Bampton, Horse Fair, **167**
Barges, 129, **136**, 137, 138
Barns, 30, 40, **43, 51**
Basket-making, **87**, 95, 139, **233**
Bees, 110, 193
Bells and Bell-ringing, 12, 175; gleaning, 194; curfew, market, pancake, 196; daily and various, 214; ringers' rules, 214
Bempton, Yorks, Egg-gathering at, 132
Best, Henry, Farm Book, 41
Bidford, **172,** 179
Bird-scaring, 64

Bishop, Election of Boy, 202; visitation of, 207
Blacksmith, **1,** 11, 84, 86
Boat and Ship-building, **128, 134, 136, 137**
" Bodgering," 117, **119**
" Bondager," 64
Bonfires, 202
Bounds, Beating the, 196, **200**
Bourne House, Bridge, **24**
Bowls, 185
Boyton Manor, Wilts, **24**
Bracken, **10,** 106
Braiding and Net-making, 96, 227
Brede Place, 29, **35**
Brightlingsea, 129, 130
Brixham Trawlers, 129
Broom-making, 94, 109–10
" Brusher " Mills, **121,** 124
Bucks, 68, 96
Burford, Oxon, **167, 184**
Button-making, 96
Buxted, Cannon-making at, 115

C

Cambridgeshire, 11, 67, 68
Canals, **9, 136,** 137, 138
Carols, Christmas, 202
Carpenter, **92,** 93, 153
Carrier and Carriers' Carts, **147,** 151, 169
Castleton, May Garland, 201
Celebrations, 31, 176, 229, 230
" Chain-rolling," 116
Chairs and Chair-leg making, 93, 95; in Bucks, 117, **119**

Chapels, private, 205 ; ruinous, 40, 205,
    210, 217 ; Nonconformist, 216
Charcoal burning, **120, 124**, 227
Charities and Doles, 80, **83**, 195, 207
Charms, 188, 203
Cheapjacks, 164
Chester-le-Street, 196
Children, 75–6, 110, 202, **77, 189, 190,
    192, 198, 200, 232** ; apprentices, 42,
    207
Chilterns, 68, 96
China clay, 109
Chipping Campden, 85, **190**
Chisel Beach, 131, 201
Christmas, mumming, **198**, 202 ; customs
    202 ; carols, 202
Church, 28, 205–6 ; registers, 206 ;
    music, 214 ; services, special, 215 ;
    210, 212, 217, 218
Churchwardens, 207 ; accounts, 206–7
Cicero on agriculture, 21
Cider, mill and press, **36**, 46 ; making,
    46
Cistercians, 69
Clamp, building the root, **62**, 64
Class Differences, 16
Clay, digging, 109
Clipping the church, 215
Clockmaker, 93
Clog-sole maker, 123
Club-walk, **172**, 176
Coach, 146, **148**
Coalbrookdale, **7**
Coast-guard, 132
Cockle gatherers, **127**, 130
Coggeshall, Essex, 69
Colchester, Oyster Feast, 195 ; winter
    custom in, 202
Combe Sydenham, Legends of, 29
Commons, uses of, 100, 106
Cooking, cottage, 73, 74
Coppice and Coppice-workers, **98**, 115,
    118, 123
Cornwall, 49, 70, 109, 130, 152, 179, 185,
    195, 196, 215, 216, 224
Corpus Christi Procession, **211**, 221
Costumes, country, 74. 75

Cottagers, **3, 4, 71, 72, 77, 78** ; earnings,
    73 ; food, 73 ; clothes, 74 ; old,
    79–80 ; character of, 81 ; livelihood,
    169
Cottages, **4**, 28, **71**, 76, **77, 78, 91, 165,
    167** ; workshops in, 90, 92, 93
Country, in the past, 2 ; difference,
    between town and, 5–6 ; diversity
    of, 11 ; occupations, 11 ; continuity
    of, 11–12 ; amusements, 15 ; pov-
    erty in, 16 ; problems of, 15–19, 235 ;
    changes in, 19, 81, 230, 235 ; statistics
    2, 235 ; future of, 20, 236
Country houses, **23, 24, 25, 26**, 28–30 ;
    Legends of, 29, **35**, 203
Countrymen, character of, 2, 5, 6, 15,
    81
Courts, forestry and manor, 112 ; baron
    and leet, 195, 196 ; of verderers,
    196 ; miners of Derby, 196
Crabbers, **128**, 130
Cream, scald, 49
Cricket, 31, 76, 180, 208
Crops, local, 68
" Crying the neck," 194
Curate, 213
Curfew, 196
Customs, general, 60, 19, 108, 131, 194 ;
    agricultural, 11, 45, 60, 63, 193–4 ;
    birth, marriage and death, 31, 193 ;
    bounds, beating of, 196 ; church,
    194–5, 215 ; Easter, 201 ; feudal,
    195 ; kern doll, 63, 194 ; May-day,
    201, **197, 190** : Shrove-tide, 196 ;
    Xmas, **200**, 202

                    D

Dairy, 49 ; — maid, 42, **38** ; — man, 11 ;
    48
Dancing, general, 176–186, 229 ; May-
    pole, Chipping Campden, **190** ; round
    Baal fire, **192**, 202 ; horn dance,
    Abbots Bromley, **191**, 201 ; hobby
    horse, Padstow, **197**, 201 ; Morris
    dance, **198, 199**, 201 ; Furry dance,
    Helston, 201

Dartmoor, 100, 105, 107; ponies **101**
Dealers, Itinerant, 144–5, 164, **156**
Derbyshire, 109, 194
Devon, 11, 40, 49, 56, 68, 109, 152
District Nurse, 19
Ditchers and Ditching, **18**, 153
Doctor, 19; country, 33; quack, 164
Donkey, 100, **119, 150, 152,** 154
Downs, 105, 230
Drake, Sir Francis, 29
Dredging, **133,** 138

E

East Hendred, Berks., 165
Easter Customs, 201
Eastington, 136
Education, Rural, 12
Eel-catching, 138, **127,** 185
Egg-gathering, 111, 138; at Bempton, 132
Elections, 186
Emsworth, Oyster Fishery, 130
Entertainers and Entertainments, **23, 142,** 144, 180, **226,** 228
Essex, 41, 64, 130, 137, 194
Estates, sale of, 19, 27, 33, 235; cost of upkeep of, 236
Evesham, 11, 68
Exe, River, **127,** 130
Exmoor, 100

F

Fairs, general, 179–180, **184, 177;** horse, **167,** 179; hiring, **167,** 179; mop, Bidford, **172,** 179
Fal, River, Oyster Fishery, 130
Farmers, in the past, 40–2, **48;** moorland, **10,** 106; wives, 49; and mortgages, 235
Farms and Farmhouses, **8,** 28, 29, **37,** 40, **43, 44, 47, 104,** 106; experimental, 31; long ownership and tenancies, 39; servants, 42, food in, 42
Feasts and Special Dishes, 186, 195; lamb roasting, 201; pig, **172,** 179
Ferries and Ferrymen, 137

Fêtes, Local, and Celebrations, 15, 31, 175, 176, **226**
Fiddler, Village, 144, 186, 202
Fire, rick, **9,** 15; Baal, 202; bonfire, 202; need, 202
Fisheries and Fishermen, 129, 131, **133,** 138, 139
Flail, 46, **51**
Flax and Flax-growing, 68, **226,** 227
Flint-knapping, 108
Flower growing, 68
Folk-lore, 131, 188, 202–4
Food, in farmhouses, 42; meals in field, **54,** 63, 75; cottagers, 73–4
Forestry, 112, **113,** 115–17, **121, 122, 225, 227, 234;** forest laws, 81, 112
Fuel, 106, 116, **119;** gathering of, 154
Funerals, 31, 193, 206, 216, **209**
Furze, **104,** 106

G

Gamekeeper, 110, 104; gibbet, **17**
Gardener, **3**
Gardens, 30, **35,** 76
Garland Day, 201
Gate-making and mending, 153
Ghosts, 29, 203
Gipsies, **10, 103,** 111
Gleaning, 63, 75; bell, 194
Gloucestershire, 115, 196
God's Penny, 193
Gossip, **71,** 75
Gravel Pits, 106, 107
Gunwalloe, All Souls' Day Service, 215

H

Hallaton, Hare Pie Festival, 201
Harvests, general, 11, **52, 53, 54,** 60, 63, 73, 194; bracken, **10;** moorland, 110; reed, 138, flax, **226;** thanksgiving services, 215
Hawker and Pedlar, 144–5–6, **156,** 158, 164
Hayfield and Hay-making, **8, 57, 58,** 60, 63, **232;** hay-baling, **231;** customs, 60, 194

Hearth, Farm and Cottage, 74, **78**
Heather Broom, 109, 110
Heathfield, fair, 179
Hedgers and Hedging, 12, **149**, 153
Helston, Cornwall, 196, 201
Hermits and Solitary Characters, **121**, 124, 146
Heveningham Hall, Suffolk, **25**
Hey Tor, quarries, 107
Hobbyhorse, Mayday, at Padstow, **197**, 201
Honiton, fair, 179
Hoop-making, 93, 123
Hop-growing, 46, 67 ; repairing wires, 57 ; picking, 58 ; oasthouses, **61**
Horn-blower of Bainbridge, **189**, 196 ; dance, at Abbots Bromley, **191**, 201 ; Wakeman's, at Ripon, 196
Horningsham, Wilts, Chapel at, 216, **217**
Horses and their Work, 40, 116, 137, 152 ; **113, 148** ; semi-wild, 100, **101** ; trappings of, 90 ; fairs, 163, **167,** 178; show, **173,** 186
House-wives, Cottage, 75
Hungerford, 195
Hunting, 180, 228 ; meets, 30, **26, 183** ; hound show, **181**
Hurdles and hurdle-making, **98** ; 118

I

Ice-cream Vendor, 145, 164, **171**
Immigration, Foreign, 6 ; middle-class, 33, 235
Implements and Tools, 11, 41, 46, **51, 53,** 55, 56, **57,** 60, 67, 86, 93, 117, 118, 123, 138
Industries, Varieties of Country, 11 ; of the past, 85, 96, 112–115 ; women's, 95, cottage, 95 ; of the Nomads, 111
Inns, **4,** 15, 50, 126, 137, **148,** 163, 174, 175, **177,** 179
Iron, Origin of Industry, 112, 115
Ives, St., 131, **136**

J

Jordan's, Meeting House, 216

K

Kelvedon, 69
Kent, 11, 40, 46, 64, 68, 137
Kern Doll, 63, 194
King's Lynn, 130
Kingsteignton, lamb roasting, 201
Kirton-in-Lindsey, Market Bell, 196
Knife-grinder, **87,** 145, **166**
Knutsford, 201

L

Labourers, frontispiece, 59, 73, 79, 144, 154, 229, **4, 8, 14, 18, 57, 78** ; wages of and allowances, 42 ;
Lace-making, **91,** 96
Lancashire, 67, 129
Landlord and tenant, 11, 41 ; long tenancies, 39 ; customs, 195, **199**
Landowners, 19 ; rise and fall of, 22, 27, 28 ; in commerce, 27, 32 ; patriotism of, 28 ; family customs of 31
Lawes, Sir John Bennet, 31
Laws and legislation, 19, 41, 115
Legends, 203 ; of country houses, 29
Lifeboat, 131
Lighthouse-keeper, 132
Lime-kiln, **102,** 108
Lincolnshire, 64, 67
Livestock, influence on cultivation, 63
Llanberis quarries, 108
Locks and Lock-keepers, **9,** 137
Longshoremen, 132
Lynchets, 68

M

Machinery, 40, 45, **47,** 49, **53,** 55, 56, 60, 67, 227, **231**
Madeley, Old Court House, **25,** 28
Manor Houses, 28–30 ; legends of, 29
Manors, rights of tenants, 112
Market Drayton, Market Place, **168**

Market and Market Days, 162–4, 228 ; bell, 196 ; **135, 168, 171**
Marshman, 138
May-day, **200,** 201 ; May-poles, **190 ;** hobbyhorse, **197 ;** superstitions of, 203
Meals, free and snacks, 195
Midsummer Day, **192,** 202
Milk and Milking, 45, **48,** 49 ; — maid, **38 ;** ewes, 41 ; boy, 75
Mills and Miller, 93, 138, 227 ; saw, **101,** 106, 115, 138 : Miller's cart, **166**
Minehead, Hobbyhorse, 201
Mining, 109
Minister, Nonconformist, 216, 229
Montacute, Gardens, **35**
Monuments and Tombs, 205, **212, 219**
Moors and Moorlands, **10,** 100, 105 ; work on, 106 ; — farm, **104 ;** markets, 163
Morals, Country, 19, 81
Morecambe Bay, 130
Morris Dancing, **198, 199,** 201
Mothering Sunday, 196
Motor bus, 151, 230
Mumming, **198,** 201–2

N

Need Fire, 202
Net-making, 96, 227
Nets, Seine, 130 ; stake and keddle, 130
Newbiggin, Northumberland, 196
New Forest, 100, **120, 121,** 124, 196, **198,** 202
Newsagent, 145, 169
Norfolk, 64, 68, 196
Nurse, District, 19

O

Oakapple Day, 201
Oast Houses, 46, **61**
Old Weston, Hunts, 195
Omens, 29, 203
Onion Seller, Breton, 146, **156**
Orchards, **44, 68**

Organ, Church, 214
Outings and Excursions, 175, 187 ; church, 215, **218**
Oxen, draught, 41, **37,** 56
Oysters, 130 ; Colchester feast, 195

P

Pack-horse, 152, **150**
Padstow, hobbyhorse, **197,** 201
Pageants, 31, 176, 179, 215 ; at Sydenham House, Coryton, **23**
Painswick, clipping the church, 215
Palm Sunday, 196
Parson, 207–8 ; wife, 213
Parsonages, 208
Patriotism, Local, 6, 15
Paythorne Bridge, Salmon Sunday at, 202
Peas, weighing, **52 ;** cultivation of, 64
Peat, uses of, 106
Pedlar and Hawker, 144–5, 146, **156,** 158, 164
Peppermint, 68
Pests and Vermin, 64
Pheasants, Rearing, 110
Pigeons and Pigeon-cotes, 30
Pigs, 152, **156,** 215
Pilchards, 129, 130
Plays, Mumming, **198,** 201, 202 ; sacred, 215
Ploughs and Ploughing, 6, **17,** 41, 55, **62 ;** matches, 15, **183,** 185 ; Plough Monday, 194
Poaching, 81, 90, 112, 146, 154
Pole-lathe, 117
Policeman, 146
Ponds, dew, 105 ; hammer, 115
Ponies, Dartmoor, Exmoor, &c., 100, **101,** 105
Population, Town and Country, 2, 235
Portland, Quarrymen, 108
Ports, Little, 126, 129
Postman, 145, **150**
Potato Cultivation, **65,** 67, 224, **234**
Pottery-making, **84,** 95
Poultry and Plucking parties, 49
Purbeck Quarries, 108

Q

Quarries, 107–8 ; **102, 103**
Quarter-day, Gifts to Landlord, 195

R

Rabbit Trapper, 110, **182**
Racing, 180, **181**
Raleigh, George, 28
Reaper and Binder, **53,** 55, 60
Reed Cutting, 138
Registers, Church, 206
Revels and Wakes, 179, 194, 215
Ripon, Horn-blowing Custom, 196
Roads and Roadside Characters, **142,**
   143, 144 ; road-mender, **149,** 152
Rogation-tide, **209,** 215
Rolling Wheat, **53**
Roots, building a clamp, **62,** 63, 64
Romney Marsh, 6
Rural Problems, 15, 16, 19, 225–6
Rush-bearing, 194
Rye, 130

S

Saddler, 90
Saintbury, **147**
Sales, cottage and farm, 161 ; estate,
   27, 33–4, 235 ; seasonal, 59, 116,
   162, **168**
Salmon and Salmon Netting, 130 ;
   Salmon Sunday, 202
Sawmills, **101,** 106, 115, 138
" Scowles," Gloucestershire, 115
Seamer, Northumberland, 179
Sea-weed gathering, 132
Sedgefield, 196
Sedgmoor, 139
Seed-raising, 68, 69
Selsea, Sussex, 215
Services, Church and Special, 215, 229
Settlements, Industrial, 16
Sexton, **212,** 213
Sheep, 11, 41, 63, 105, 152, 162, 194 ;
   washing, **13,** 105 ; dipping, **38** ;
   shearing, **43,** 45, 105 ; customs, 194

Shepherd, 105, 186, 230
Shenington, Oxford, 195
Sherborne, Pack Monday Fair, 179
Sherwood Forest, 112
Ships and Shipping, **128, 134, 136** ;
   building, 137 ; coasting, 126
Shoe-maker, 90, **92**
Shops and Shop-keeping, **18,** 159–161,
   165, 228
Shooting, 30, 50, 138, **178,** 180, **182**
Shows, agricultural, 180 ; hound, **181** ;
   horticultural, 186
Shrovetide, customs, 196
Smock-frock, 74
Society in the Village, 16, 32
Somerset, 40, 68
" Souling," 202
Sowing and Planting, 56, 59, 194
Sport, 30, 31, 50 ; **178,** 180, **182** ;
   local, 185 ; country house sport,
   26, 30
Squatter, 79, 146
Stallion, Travelling, **148,** 152
" Statesmen," 39
Statistics, rural population, 2, 235 ;
   war-time cultivation, 224
Stonemason, 107
Sussex, 56, 70, 105, 115, 130, 215
Swaling, 106
Sydenham, Sir George ; Elizabeth,
   Combe, 29 ; House, **23**

T

Tailor, 89
Tanner, 90
Taunts, **91,** 131
Teasel Growing, 68
Tenants and Tenancies, 11, 39, 41, 195
Thatcher, **91,** 94, **97**
Thaxted, Essex, **7**
Threshing, 46, **47, 51**
Tinker, **87,** 93, **166**
Tissington, Derby, Well-dressing, **192,**
   195
Tithes, 208
Tramps, 143, **155**

Tree-felling, 116, **121** ; grubbing stumps, 117, **122**
Tugs, Steam, 131

V

Valentine's Day, St., 196
Vegetable Growing, 68
Verger, 213
Virgin's Crown, 193

W

Wages, 42, 73, 79, 89, 95
Waggon, **113**, 116, 146, **166**
Wales, 56, 109, 130
Warrener, 110
Wastes, Uses of 100, 106
Watercress, **66**, 68
Water Diviner, 93
Weather, Effects of, on Farming, 5, 55, 63
Wells, 79 ; dressing, **192,** 195
West Stafford, Dorset, **217**
Whalton, Northumberland, **192,** 194, 202
Wheelwright, 11, 86, **88,** 89, **233**

Wherry, 137
Whist Drives, 15, 186
Whitstable, 130
Whitsun Customs, 201
Whortleberries, 110
Wild Fowling, 138, **178**
Wills, 41
Winter, Announcing the Arrival of, 202
Wirksworth, Derbyshire, **189**
Withy Growing, 139
Woad, 68
Women and Women's Work, **3,** 64, 70, 71, 74, 75, **77, 78, 83,** 95, 224, **225, 226,** 227, **231, 232, 233, 234**
Wooden Ware, making of, 93, 94
Workshops, Cottage, 90, 92, 93.
Wroth Money, 195

Y

Yacht, 131, 138
Yorkshire, 6, 41, 67, 137

Z

Zennor, 131

www.ingramcontent.com/pod-product-compliance
Lightning Source LLC
Chambersburg PA
CBHW061724270326
41928CB00011B/2104